# A PHOTOGRAPHIC TOUR OF THE UNIVERSE

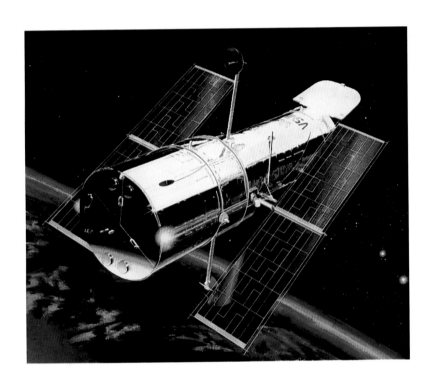

# Gabriele Vanin

# A PHOTOGRAPHIC TOUR OF THE UNIVERSE

With a Foreword by Richard M. West

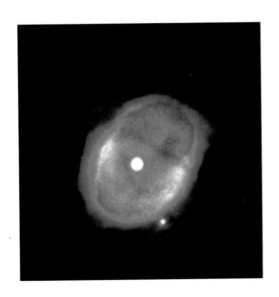

FIREFLY BOOKS LTD.

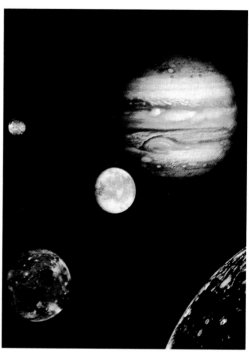

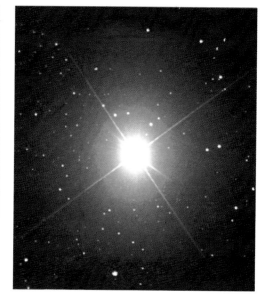

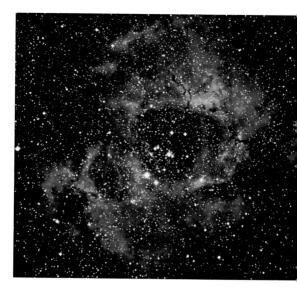

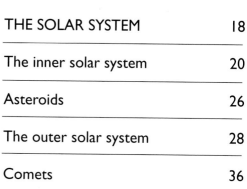

A FIREFLY BOOK

Published by Firefly Books Ltd. 1999

Copyright © 1998 Arnoldo Mondadori Editore S.p.A., Milan

All rights reserved. No part of this publication may be reproduced, stored in a retrieval system or transmitted in any form or by any means, electronic, mechanical, photocopying, recording or otherwise, without the written permission of the publisher.

Third printing 2000

**Cataloguing in Publication Data**

Vanin, Gabriele.
    A photographic tour of the universe

Rev. & expanded ed.
Includes index.
ISBN 1-55209-345-X

1. Astronomy - Pictorial works.
2. Astronomical photography. I. Title.

QB68.V36 1999        520' .22'2     C99-932739-6

Published in Canada in 1999 by
Firefly Books Ltd.
3680 Victoria Park Avenue
Willowdale, Ontario
M2H 3K1

Published in the United States in 1999 by
Firefly Books (U.S.) Inc.
P.O. Box 1338, Ellicot Station
Buffalo, New York
14205

Printed and bound in Spain
D.L. TO: 588-2000

# Contents

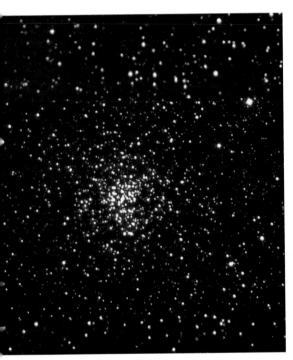

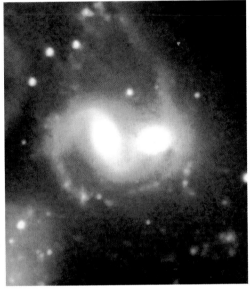

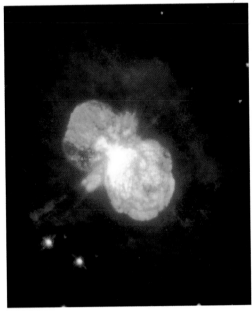

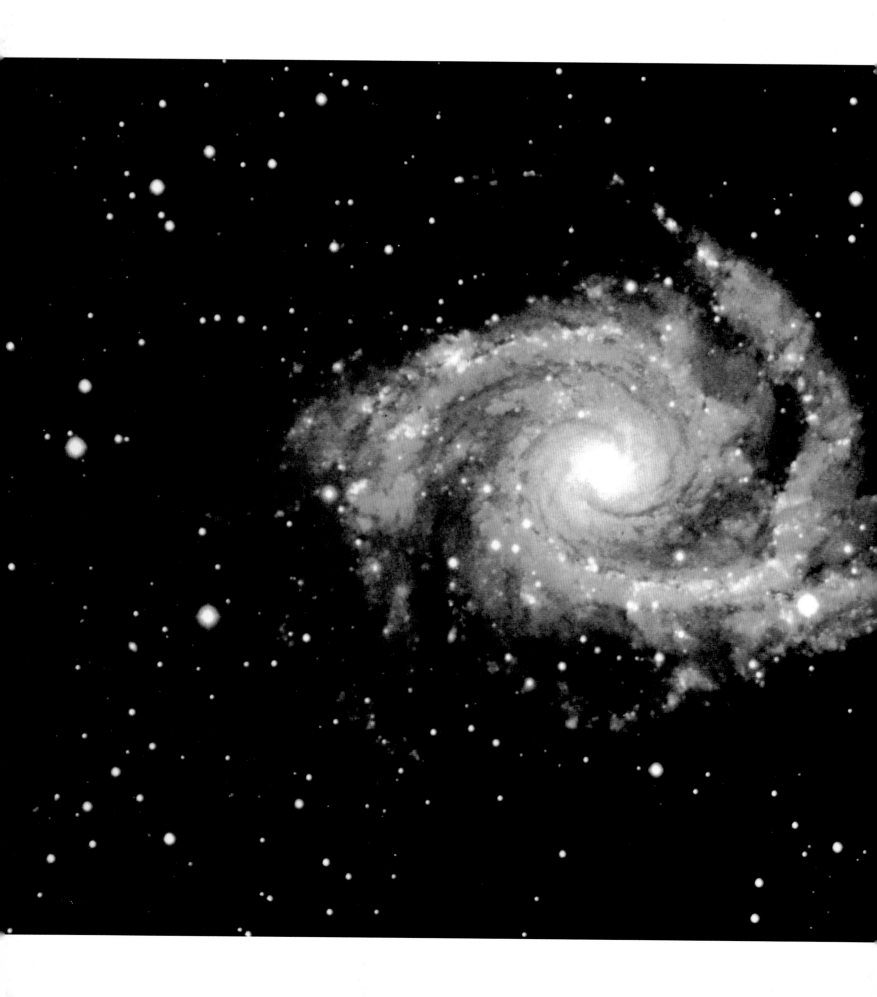

# Foreword

Astronomy may well be the oldest science, but it is also a very modern one. Many centuries have passed since humankind first began to study the heavens, and it has taken a long time to reach the highly advanced stage of contemporary astronomy and astrophysics. All of this has been driven by our primeval quest to understand better the mysterious space that surrounds us, to learn more about our distant origins and to comprehend more clearly the limitations of our small niche in time and space. Guided by our natural curiosity, combined with our ability to construct better and better instruments, we have progressively penetrated deeper and deeper into the universe and its many riddles.

Astronomy is an exploratory science through which we all may undertake incredible voyages into the unknown. It also has the particular advantage of producing spectacular, colorful images. Although we may not yet fully understand the physical realities behind the celestial objects they depict, we may certainly admire their beauty and marvel at their complex and intricate structures. And we enjoy them best of all when the stories behind these images are told, when they are explained to us in an authoritative and interesting manner.

This is the principal achievement of the present book. It presents a marvelous collection of the latest, most significant astronomical images, obtained with the largest, most powerful telescopes now in existence. Extensive captions point out the details of each picture, how it was made and its astronomical significance.

The images are complemented by a comprehensive text that begins with the basics of astronomy, describing the various instruments with which astronomical observations are made. Next, we are taken on an exploration of the solar system, then outward to the stars and nebulas and, finally, to distant galaxies and quasars located in remote regions of the immense universe. Along the way, we learn about important historical developments and the latest discoveries, from the newly found transneptunian objects to the fantastic images of newborn stars and multi-million-year-old galaxies recently obtained with the refurbished Hubble Space Telescope.

The author—a central and respected figure in Italian amateur astronomy—and the publishers are to be congratulated on this excellent book. It has surely not been a simple task to collect all of these images from so many different sources, but the result is outstanding. While offering valuable information to the initiated, this volume also has a great potential to attract many newcomers to this wonderful science. I am sure its readers will spend many pleasant and instructive hours in its company!

Richard M. West (European Southern Observatory)

# The colors of the sky.

What we know is, to a large extent, first and foremost what we see. This is true in all fields, even in science, and especially in that most visual of sciences, astronomy. Vision is worth nothing if it is not supported and corroborated by reasoning and interpretation, but there would be nothing to discuss if we had no perception of the nature of the universe. This perception occurs above all by means of our most powerful sense, the sense of sight. For millennia, astronomy was practiced solely with the naked eye. There is evidence of observations of

the sky dating from the Paleolithic period, more than 10,000 years ago. Even the Babylonians, 4,000 years ago, achieved a considerable degree of sophistication in observing and predicting the positions of celestial bodies. Later, the Greeks, as far back as 2,500 years ago, succeeded in developing complex models of the cosmos and also came to understand such concepts as the shape and size of the Earth, the dimensions and distance of the Moon, the relationship between the dimensions of the planetary orbits and the nature of the Milky Way.

Our eyes, however, are organs that are limited in terms of their sensitivity to the spectrum of light. This means that we cannot see objects that are too poorly lit or that lie outside the region of electromagnetic radiation known conventionally as visible light.

## The advent of the telescope.

A first aid to increasing the power of the human eye came in the early years of the 17th century with the invention of the telescope, which Galileo Galilei was among the first to turn to the sky in 1609. Viewed with skepticism in the beginning, the new instrument soon

opened up the doors of the sky to astronomers and contributed decisively to completing the reform of astronomy begun by Copernicus a few decades earlier. By the end of the 17th century, the classic refracting telescope (with lenses) was joined by the reflecting telescope, invented by Issac Newton in 1671. Its basic concept was that an image was formed by reflection of light onto a parabolic mirror placed at the bottom of the instrument. The most important instruments (though not always the most perfect) of the 18th and 19th centuries (and indeed also of the present day) were reflectors. One need only think of the 122-cm telescope built by William Herschel in 1789 and Lord Rosse's 183-cm "Leviathan" (1844).

Such instruments enabled constantly improving observations of the fainter and more distant celestial bodies. Visual observations, however, could at that time be recorded only by means of written notes and sketches. Distracted astronomers often failed to keep notes, and many discoveries, such as that of the planet Uranus, were delayed because of this bad habit. In addition, drawings tended to be largely subjective

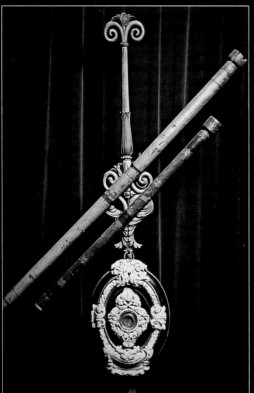

Above, two of the telescopes built by the great Italian physicist and astronomer Galileo Galilei, preserved at the Museo di Storia della Scienza, in Florence. The one on top has a focal length of 1,330 mm, an effective aperture of 26 mm and a magnification of 14x. The one at the bottom has a focal length of 880 mm, an effective aperture of 16 mm and a magnification of 21x.

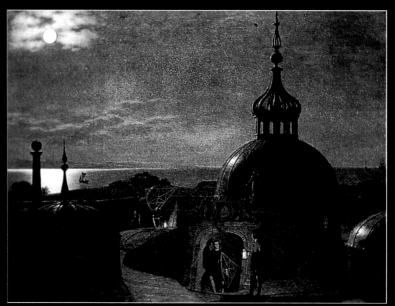

Left, Uraniborg, the observatory of Tycho Brahe (1546-1601) on the island of Hven, in Denmark. Brahe, who died just 8 years before Galileo turned his telescope to the sky, was the last great naked-eye observer and undoubtedly the greatest of all. His instruments were a quadrant and armillary spheres; nevertheless, he was able to carry out observations with them that were 10 times more accurate than those of his contemporaries.

and unreliable, especially if the astronomer could draw only a little or drew badly.

A wonderful tool for the objective recording of observations soon came to the aid of astronomy—photography. The first experiments using the camera obscura go back to the ninth-century Arab astronomer Al-Kindi, and those employing the photochemical action of light date from as far back as Aristotle. In 1566, Giorgius Fabricius observed the darkening of silver chloride, and in 1727, J.H. Schultze demonstrated that this was due to light. In the early 19th century, English chemist Thomas Wedgwood obtained pictures of small objects, though he was unable to fix them in a permanent way. In 1810, Thomas Seebeck succeeded in capturing on a sheet of paper covered with damp silver chloride a few colors of the spectrum produced by a prism through which light was passed. The French inventor Joseph Nicéphore Niepce, in 1822, was the first to fix an image (a still life) permanently on a glass plate covered with bitumen of Judea.

Around 1820, the painter Louis Jacques Mandé Daguerre tried to record the colors of objects by using phosphorescent minerals. Daguerre and Niepce began working together, exposing silver-plated copper plates to iodine vapors. After Niepce's death, Daguerre completed the work in 1835. The image formed on the plate was revealed by exposing it to mercury vapors, which amalgamated with the silver of the

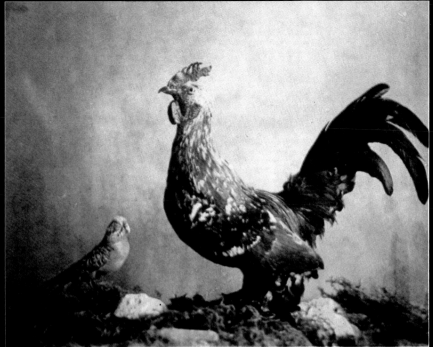

One of the first color photographs. It was taken by Louis Ducos du Hauron by combining negatives taken through blue, green and red filters.

plate's exposed areas. Thus, the daguerreotype was invented, though Daguerre kept it secret until 1839. The procedure, which produced positive images directly, came to be perfected in the years that followed. Meanwhile, another French experimenter, Hippolyte Bayard (who has been forgotten by the official histories), succeeded in obtaining positive images on paper. Also in the year 1839, English astronomer John Herschel invented what we can truly call photography—the process of forming a negative image on a surface that darkens when exposed to light, fixing the image and printing a positive image from it onto a second surface, similar to the first. Herschel used ordinary paper covered with silver nitrate and, as a fixing agent for the image, sodium hyposul-

fite. The same procedure had been discovered independently by the physicist W.H.F. Talbot (who named it calotype), but without an adequate permanent fixing agent. This was the precursor of the modern photographic process.

Herschel magnanimously renounced any credit for the original discovery in favor of Talbot, but it was he who coined the terms "photograph," "negative" and "positive." In 1840, Herschel succeeded in repeating Seebeck's experiments

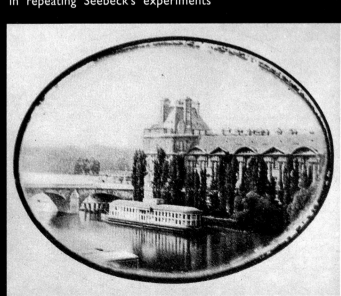

One of the first daguerreotypes: a view of the Tuileries Palace produced on silver plate by Louis Jacques Mandé Daguerre in 1839 (Paris, Museum of Arts and Crafts).

with color reproduction, and in 1847, French physicist Alexandre Edmond Becquerel registered the colors of the spectrum on a layer of silver chloride stretched over a daguerreotype plate, instead of on paper. In 1851, Claude Niepce de Saint Victor, nephew of Joseph Niepce, managed to improve on Becquerel's procedure by making the colors more permanent with a protective varnish. However, it was still not possible to fix color pictures permanently, and the exposure time for recording all the colors was too lengthy, taking up to 1 hour.

### Fixed stars.

In the meantime, the sky had begun to be a subject for photography. The first attempts were made by Daguerre himself, at the suggestion of astronomer François Arago, and even predate the public announcement of the daguerreotype. The target was the brightest of all night-sky celestial bodies, the Moon; but the attempt failed. The first successful daguerreotype of the Moon was produced in March 1840 by John Draper of New York, using a 20-minute exposure, but it was subsequently destroyed in a fire. In 1841, Draper succeeded in producing the first daguerreotype of the solar spectrum. The oldest surviving da-

guerreotype of the Moon dates from September 1, 1849, and is the work of another New Yorker, S.D. Humprey, who used a 200-mm lens. In 1845, at the Paris Observatory, Armand Fizeau and Léon Foucault took the first picture of the Sun, on which a few sunspots could be seen, with an exposure time of one-sixtieth of a second.

The first photograph of a star, Vega, dates from July 16, 1850. It was immortalized in a 100-second exposure taken with the 38-cm refractor at the Harvard College Observatory by the director, G.P. Bond, and photographer J.A. Whipple.

The earliest photographs of a solar eclipse date from 1851 and were taken by Berkowski at the Königsberg Observatory in Prussia, using a 16-cm telescope. In 1852, England's Warren De la Rue was the first to employ the collodion "wet plate" process in astronomy. The technique was developed by Frederick Scott Archer, also of England. In this process, potassium iodide was added to a glass plate covered with collodion, and the whole was then immersed in a solution of silver nitrate. Using a 33-cm reflecting telescope, De la Rue took pictures of the Moon, with exposures of 10 to 30 seconds, showing

its main features. From 1858, he began to take daily pictures of the Sun, and in 1860, he organized an expedition to Spain to observe a total eclipse, during which solar prominences were recorded for the first time. The recording of the solar corona had to wait until the 1870 eclipse, when Brothers succeeded in photographing it from Syracuse.

In December 1864, American astronomer Lewis Morris Rutherfurd built the first refracting telescope that was specifically designed for astrophotography (29 cm in diameter). Unlike reflectors, conventional refractors could not make all the colors of the spectrum converge at a single point. This new instrument enabled Rutherfurd to take the first photograph of a star cluster, the Beehive Cluster, with a 3-minute exposure. In just 13 years, using a larger photographic and visual refractor with a 33-cm aperture, he produced no less than 1,400 photographs of the sky, 650 of them pictures of star clusters and groups, including more than 50 photographs of the Pleiades. Rutherfurd used the collodion wet plate process, which had by then overtaken both calotype and daguerreotype. Collodion plates were much faster, making exposures an average of 10 times shorter. However, the plate had to be used as soon as it had been prepared, before the collodion dried out, which meant that an exposure could not be very long. Rutherfurd generally did not exceed 6 minutes.

In 1858, the first diffuse object was cap-

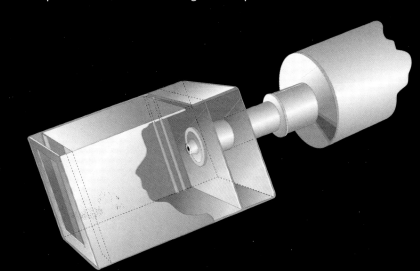

Below right, how early cameras were fitted to the eyepiece of a telescope in the pioneering days of astrophotography. The camera was a simple camera obscura, a light-proof box. The lens consisted of the telescope's objective lens and eyepiece. The shutter was inserted manually into the opening which is visible just behind the eyepiece. The photographic plate was placed inside the slit visible at the bottom of the box.

tured—the spectacular Comet Donati. English photographer Usherwood captured its image using a simple portrait lens (the comet was very large) and just 7 seconds' exposure time, but it showed only the coma and the beginning of the tail. Effective pictures of diffuse objects had to await a further innovation: the introduction of dry plates, made up of silver bromide gelatin. These could be used some time after preparation and were also much more sensitive than collodion plates. Using these, John Draper's son Henry took the first photograph of the Orion Nebula (see page 61), and in 1881, Henry Draper, together with Jensen, successfully photographed Comet Tebbutt. The first beautiful photograph of a galaxy, An-

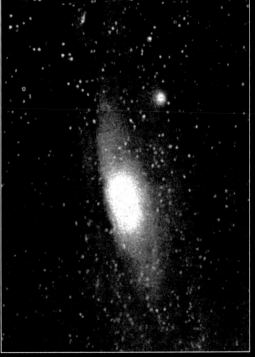

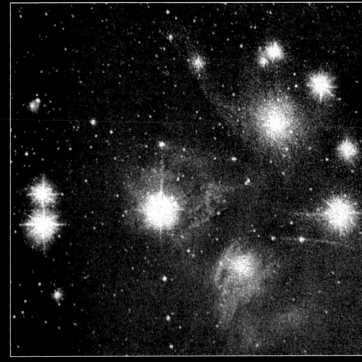

dromeda, dates from 1888 and was taken by American astronomer Isaac Roberts with a 51-cm reflecting telescope and a staggering 4 hours' exposure. Astronomical photography had finally established itself and came to replace normal sky-survey techniques: a single 30-minute exposure could save a week's work at the drawing board. Although film is not much more sensitive than the human eye, it has the advantage of being able to accumulate light over a long period of time, enabling a vastly improved view of celestial objects.

**Above, two early astronomical photographs by Isaac Roberts: The Pleiades star cluster (right), a 6-hour exposure taken in 1886, and the Andromeda Galaxy (left), a 4-hour exposure taken in 1888.**

In 1881, French physicist Gabriel Lippmann produced the first direct, permanent color photographs, using glass plates covered with silver bromide placed inside the camera in contact with a layer of mercury. The colors were visible, however, only by ob-

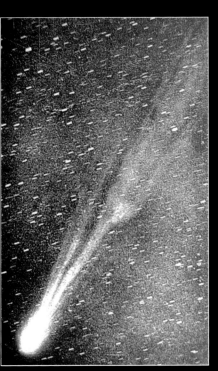

**Above, Comet Swift photographed by Edward Emerson Barnard in 1892.**

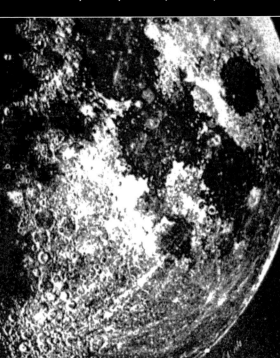

**Left, a picture of the Moon taken by Lewis Morris Rutherfurd in 1865.**

serving the exposed plates at a certain angle. Modern processes of color photography are based on the theory of color vision put forward as far back as 1801 by English physicist Thomas Young.

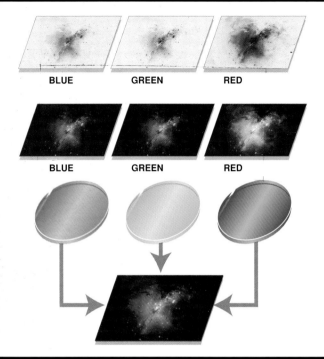

BLUE  GREEN  RED

BLUE  GREEN  RED

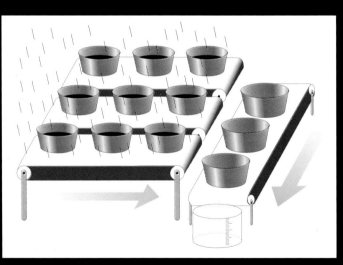

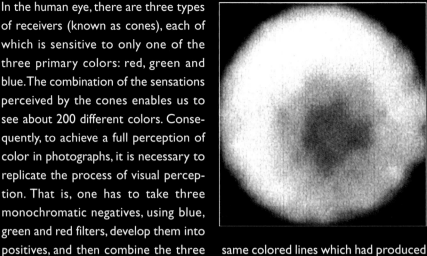

In the human eye, there are three types of receivers (known as cones), each of which is sensitive to only one of the three primary colors: red, green and blue. The combination of the sensations perceived by the cones enables us to see about 200 different colors. Consequently, to achieve a full perception of color in photographs, it is necessary to replicate the process of visual perception. That is, one has to take three monochromatic negatives, using blue, green and red filters, develop them into positives, and then combine the three positives by means of either an "additive" or a "subtractive" process. Additive methods reconstruct images by mixing the light beams of the three primary colors; for example, by superimposing three positives projected onto a screen, as English physicist James Clerk Maxwell did for the first time in 1861. He used the same filters for projecting the image that he had used to take the negatives. In 1869, France's Louis Ducos du Hauron succeeded in printing onto paper the first images obtained by superimposing three color positives. He subsequently devised a method that enabled the use of a single emulsion exposed by means of a glass screen on which were traced very fine lines of the three primary colors. Each of the lines was transparent only to light of the corresponding color. When a second screen similar to the previous one was superimposed onto the positive obtained in this way, the exposed areas of the positive reflected light through the

same colored lines which had produced the image and which, blending in the eye of the observer, reproduced the colors of the original. Various processes were developed from Ducos du Hauron's idea, but the only one to have any commercial success was the one known as Autochrome, created in 1907 by the Lumière brothers, in which the colored screens were replaced by filters consisting of colored grains of starch superimposed on the emulsion.

Beginning in 1874, the dynamic Ducos du Hauron also experimented with subtractive processes. In a subtractive process, the recombining of the positives is accomplished with color dyes—cyan (blue), magenta and yellow—that are complementary to the primary colors. Each selectively absorbs one of the primary colors (red, green and blue, respectively), allowing perception of the other two. In 1897, Ducos du Hauron first conceived of the "tripack," which combines three different color-sensitive emulsions on a single base. The ma-

Top left, the three-color, additive process used by David Malin for composing his magnificent color photographs. Above right, what is believed to be the first astronomical picture obtained by means of a CCD camera: the planet Uranus photographed in 1975 in near-infrared, at a wavelength of 890 nm, with the 155-cm reflecting telescope in the Santa Catalina Mountains near Tucson, Arizona, by astronomers of the Jet Propulsion Laboratory and the University of Arizona. Bottom left, CCDs function in a manner like that of collecting rainwater in a series of receptacles placed on conveyor belts. In this case, the receptacles are filled with light. As each one reaches the end of the belt, it is emptied into another receptacle placed, in turn, on a belt that carries it until it reaches the electronic monitoring device at the end, which creates an image from the accumulated light.

terials that are used in color photography today are refinements of this process.

## The sky in color.

Curiously, no one thought to use color photography for astronomy until 1959, perhaps because it was not considered to be scientifically useful. In that year, William Miller began to photograph various celestial objects with the telescopes on Mount Palomar and Mount Wilson, using commercial films such as High Speed Ektachrome and Super Anscochrome. The pictures obtained were of excellent quality. Seven of them are reproduced in this book, in chapters 3, 4 and 5. The use of a single commercial emulsion, however, is not particularly satisfactory in astronomy: with the long exposures that are necessary, the film declines dramatically both in sensitivity and in color reproduction. It is therefore necessary to return to the origins, in a sense: making three negatives using filters and films sensitive to the three fundamental colors, then combining them with the subtractive or additive technique (the three-color process known as tricolor). One such method, the Kodak dye transfer, has been employed by various photographers, including Joseph Miller at the Lick Observatory (for photographing nebulas) and James Wray at the McDonald Observatory (for galaxies), with reasonable results. These results have, among other things, shown how color photographs can be scientifically useful; for example, for revealing areas of star

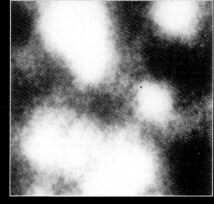
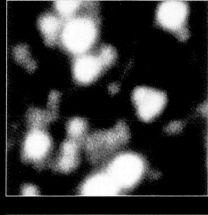
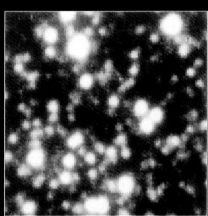
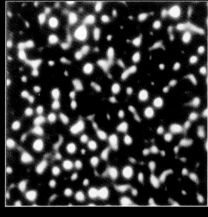

Left, four images of the center of the globular cluster Omega Centauri. The picture at top left was taken with the 1-m Schmidt telescope at the European Southern Observatory (ESO) at La Silla, Chile. The picture at top right was taken with the ESO's 3.6-m telescope, using traditional optics. Compare these with the two shots below them, taken with the ESO's 3.6-m New Technology Telescope (NTT), which uses an active optics system. Note that the NTT image at lower left is grainy, whereas the one on the right has been enhanced to improve the resolution. The four shots have respective resolutions of 2.0, 1.0, 0.33 and 0.18 arc-second.

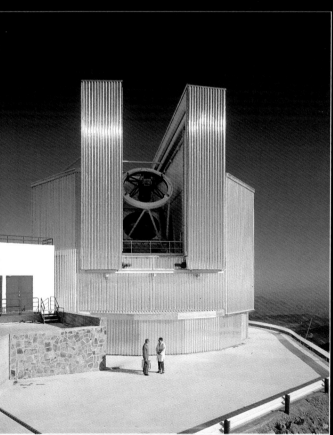

Above, a diagram showing how the active optics of the NTT (right) work: light reflected from a mirror positioned in front of the telescope's focus is analyzed by a device connected to a computer, which transmits the necessary corrections to the actuators placed under the primary mirror.
1) Telescope's focus
2) Image analyzer
3) Computer
4) Mechanical actuators

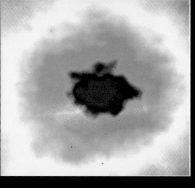

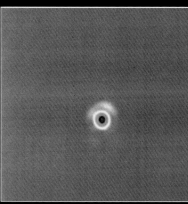

**On this page, a few examples of the performance of adaptive optics. Above, false-color images of Capella, a star in the constellation Auriga, taken by Robert Fugate at a wavelength of 890 nm (near-infrared), with the 1.5-m telescope at the Starfire Optical Range in New Mexico. The top picture, not corrected, appears very blurred, with a resolution no better than 2 seconds of arc. The lower picture, produced with a 10-millisecond exposure and an adaptive correction made with the help of a laser beam, achieves a spectacular resolution of 0.13 arc-second.**

formation, stars of different populations, different types of nebula, details in the structure of the gas in supernova remnants, degrees of ionization of the gas in planetary nebulas, and so on.

The credit for having suggested color astrophotography as a formidable means of scientific enquiry goes to David Malin, a British professional photographer who, working at the Anglo-Australian Observatory in New South Wales, Australia, has produced, between 1977 and the present, nearly 200 photographs that are both spectacular and scientifically valuable (25 are reproduced in this book). Malin uses the venerable additive technique pioneered by Maxwell more than 130 years ago. He takes three black-and-white pictures of the same subject, using a combination of plates (hypersensitized in solutions of nitrogen and hydrogen in order to increase their speed) and filters suitable for recording images in blue, green and red light. Sometimes, he uses plates from images taken by his colleagues. From the negatives, Malin derives three black-and-white positives which are then projected together through an enlarger and three filters of corresponding color onto a large sheet of color film. From the negative thus produced, he then creates the final print, employing special processes where necessary to show extremely faint objects (photographic amplification) or to modulate the considerable variation in brightness among the components of a celestial object (unsharp masking).

Today, most color astronomical pictures are taken by amateurs. Few use the tricolor process, whereas many take two commercial-film negatives of the same object, then superimpose and print them together in positive to emphasize the color and visibility of fainter details. They may also employ two additional procedures, duplicating the negatives onto transparencies and, if necessary, also on inter-negatives. The masters of this technique are American astrophotographers Tony and Daphne Hallas (three of their photographs are reproduced in this book). The Japanese astrophotographer Akira Fujii, one of the most admired in the world, uses traditional techniques (three of his pictures can be found in this book).

## The role of electronics.

In the late 1970s, chemical photography began to be replaced by electronic photography using a device known as a CCD (charge-coupled device), invented in 1970 by Willard Boyle and George Smith of Bell Laboratories. In CCD photography, light is transformed into electrical charges; the light particles, or photons, which enter the telescope, hit a thin semiconductor silicon chip (a CCD) and displace electrons, thus producing alternate negative and positive charges. These move in sequence along the chip and are collected and measured at the end. Taking pictures with a CCD camera enables one to collect most of the incident light, whereas tra-

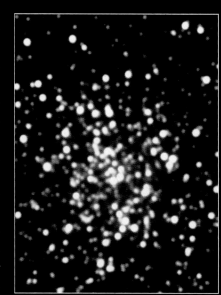

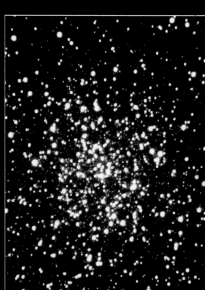

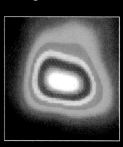

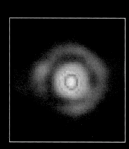

**Left, two images of the star HR 6658 taken with the 3.6-m ESO telescope. Adaptive correction improves the resolution from 0.75 arc-second (far left) to 0.22 arc-second (near left).**

**Above, the center of the globular cluster M3 photographed with the CFHT by Hesser and Bolte. The High Resolution Camera, an adaptive optics system, improves resolution from 1.2 (left) to 0.44 arc-second (right).**

ditional photography records only approximately 1 percent of it. This improved efficiency makes it possible to reduce the exposure time tenfold, from hours to mere minutes.

CCDs can even take color pictures by means of an electronic tricolor process; that is, using a computer to combine three separate shots filtered through the three primary colors. Chapter 5 presents several examples of this technique, which is widely used at the Kitt Peak observatory. Chapter 6 shows the most spectacular results yet obtained—images taken with the Wide Field/Planetary Camera on the Hubble Space Telescope. Despite their speed and immense versatility, however, the small size of CCD chips means they are able to take pictures only in a somewhat restricted field, as compared with the photographic plate (used mainly with the large Schmidt photographic telescopes, which have an enormous field). Another factor, at least up to now, is that CCDs cannot equal traditional photography's ability to record particular moments and to render colors accurately over wide areas. The reader can see this from the pictures shown.

Celestial objects are currently studied not only in visible light but also in other wavelengths of radiation. As far back as 1801, W. Herschel and J. Ritter demonstrated the existence of luminous types of radiation with wavelengths stretching beyond the visual spectrum—beyond red (infrared) and beyond violet (ultraviolet). In 1887, Hertz discovered radio waves, and in 1895, Roentgen discovered X-rays. In 1931, Karl Jansky discovered the first radio waves coming from space, and in 1948, X-rays from space were observed for the first time. In 1969, the first celestial source of gamma rays was identified.

The sky can be studied from the Earth's surface in radio waves and also, to some extent, in infrared and ultraviolet. But these last two portions of the spectrum, as well as the entire gamma ray and X-ray regions, can be observed fully only in outer space. In 1946, a V-2 rocket launched from the United States captured the first ultraviolet solar spectrum. Since then, there have been innu-

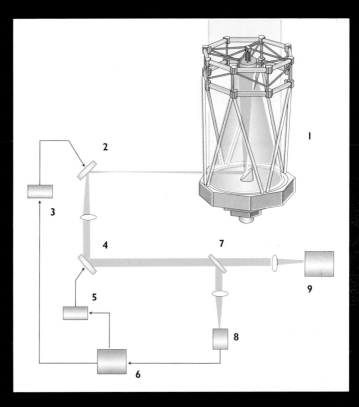

merable rocket and satellite launches that have investigated the sky across the entire electromagnetic spectrum. In this

Above, a diagram showing how adaptive optics work when applied to an astronomical telescope to correct the image of a celestial object distorted by atmospheric turbulence.
1) Telescope
2) Correcting mirror
3) Motor
4) Deformable mirror
5) High-voltage motors
6) High-velocity processor
7) Optical beam splitter
8) Wave-front sensor
9) Camera

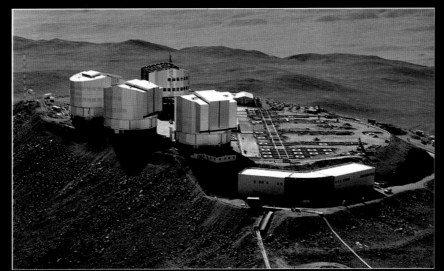

Left, two adaptive-correction lasers converge in the constellation Draco, at which the 1.5-m telescope at the Starfire Optical Range in New Mexico is pointing (photograph by Roger Ressmeyer).

Above, an aerial view of Cerro Paranal in Chile shows buildings set to house the European Southern Observatory's Very Large Telescope (VLT), four 8-m mirrors that together will function as the equivalent of a gigantic 16-m telescope.

Below, a compelling panoramic view of the European Southern Observatory at La Silla, Chile. With its 15 domes housing telescopes ranging in diameter from 40 cm to 3.6 m, this is one of the most important astronomical centers in the world.

book, the reader will find pictures of celestial objects taken in all regions of the spectrum. Going beyond the Earth's atmosphere is useful not only to observe in wavelengths other than that of visible light but also to overcome problems connected with atmospheric turbulence that affect optical telescopes. The atmosphere is anything but transparent; on the contrary, it is disturbed by all kinds of currents that ruin astronomical pictures and diminish their quality at least tenfold. It was for this reason that an orbiting space telescope was proposed as far back as the 1960s, something that has only recently become a working reality. The Hubble Space Telescope "sees" 10 to 20 times more clearly than terrestrial telescopes, resolving celestial details down to 0.05 arc-second, which is equivalent to 1/37,000 of the apparent diameter of the Moon as seen from the Earth. This is like seeing a small coin at a distance of 100 km.

Still, the good news for ground-based telescopes is that it has recently become possible to "correct" images distorted by atmospheric turbulence by employing complex electronic mechanisms known as "active optics" and the even more complex "adaptive optics." Active optics systems compensate for the effects of atmospheric turbulence, environmental fluctuations and changes in the telescope's orientation by making slight adjustments to the telescope's mirrors.

In adaptive optics systems, the image is analyzed hundreds of times per second by computers that monitor the effects of atmospheric turbulence and then are able to correct an image by using actuators that deform the surface of a telescope's mirror sufficiently to compensate for the atmospheric distortion. Adaptive optics were first suggested by Horace Babcock in 1953. His work in this area, however, was forgotten until the early 1980s, when the subject acquired a military significance within the United States Strategic Defense Initiative (SDI), otherwise known as the "space shield."

To gauge the amount of distortion being created in a telescope's image by atmospheric turbulence, adaptive optics systems need a comparative reference point, such as a bright, known star. However, it is rare to find sufficiently bright stars in the field of an interesting astronomical subject. For this reason, studies are being done on the possibility of creating an "artificial star" by means of a laser beam projected into space. Since

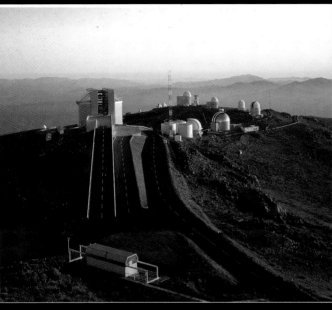

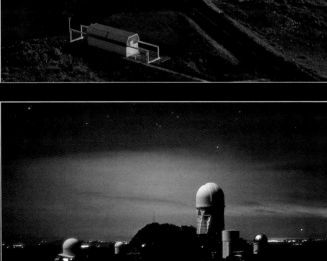

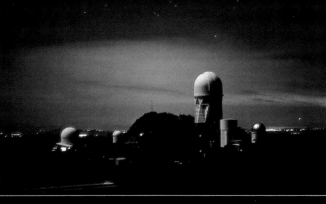

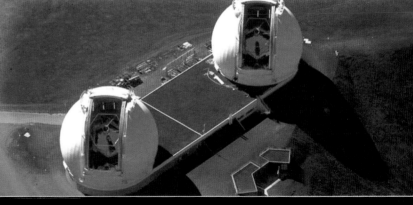

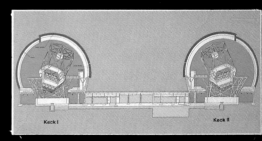

Above, the domes of Kitt Peak National Observatory, 90 km from Tucson, Arizona, illuminated by the Moon. Above right, the Keck Telescopes—the world's biggest telescope pair—each have a 9.8-m-diameter mirror. They sit on the summit of Hawaii's Mauna Kea. At 4,150 m, this is the world's best astronomical site in terms of sky transparency, calm atmosphere, number of clear nights per year and distance from light pollution. At right, diagrams of the instruments show their segmented structure. Each is composed of 36 hexagonal 1.8-m mirror segments functioning together and kept in alignment by a sophisticated system of electronic controls based on active optics.

Keck I          Keck II

research carried out within the context of the SDI was declassified in May 1991, it has become known that the Pentagon developed laser-guided adaptive optics for the destruction of missiles as far back as October 1981. The technique is based on the reflection of a laser beam by a layer of sodium atoms unleashed at a height of 90 km, which creates a kind of "lighthouse beacon," or "artificial star," to enable calibration of the optical system. One laser alone is not enough, though, because it can correct only a small part of the visual field; it is necessary to employ a whole array of lasers to obtain a sufficiently wide field.

Various experiments in adaptive optics have been carried out in recent years, with very good results. These demonstrate that there is no theoretical barrier to compensating for the effects of atmospheric turbulence in terrestrial telescopes. For 1-to-2-m telescopes, the anticipated theoretical resolutions with adaptive optics have already been reached—down to 0.1 arc-second, which is 5 to 10 times better than for images that have not been corrected. It is hoped that soon, the same performance will be achieved even by 4-m telescopes. If so, these instruments will be able to overtake the results of the Hubble Space Telescope, achieving a resolution of 0.03 arc-second. However, it must be pointed out that no adaptive optics system has yet succeeded in producing consistent results anywhere near those of the Hubble. The best adaptive optics working full-time on

large instruments are set up on the William Herschel Telescope on the Canary Islands and the 3.6-m Canada-France-Hawaii Telescope (CFHT) in Hawaii, and these generally achieve results that are no better than those of good active optics, that is, about 0.35 arc-second. Very soon, however, two telescopes of similar size to these will be in a position to improve on this.

One is the 3.5-m reflector at the Starfire Optical Range in New Mexico; the other is the Italian-built 3.6-m Galileo National Telescope in the Canary Islands, which commenced operating in 1996.

For the moment, adaptive optics can complement the Hubble Space Telescope, rather than compete with it. The future of optical astronomy, though, appears to lie on Earth: an adaptive optics telescope costs, in terms of diameter, just 1 percent of the cost of a space

telescope. In 2001, the European Southern Observatory's Very Large Telescope (VLT) will be completed in Chile (the first of its four telescope units is already operational). With its equivalent 16-m diameter, the VLT will have adaptive optics that could break the threshold of 0.01 arc-second!

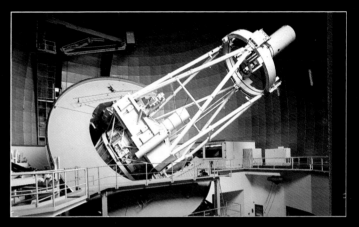

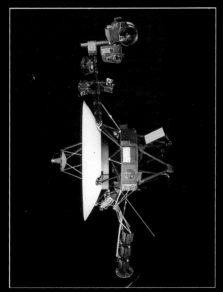

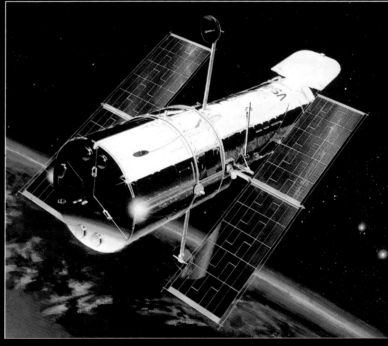

**Above, two symbols of space astronomy. Above left, the Voyager 2 probe, which went on a triumphal parade across the solar system between 1977 and 1989, when it reached Neptune. Above right, the Hubble Space Telescope, launched in April 1990 and fully corrected in December 1993, continues to yield a string of important discoveries in all fields of astronomy.**

# The solar system.

Around 4.5 billion years ago, about 30,000 light-years from the center of the Milky Way Galaxy, an area rich in dust and gas began to break up and coalesce into clouds of interstellar matter, perhaps because of a disturbance created by one or more nearby supernova explosions. As the hot, expanding gases from the stellar explosions collided with the material of the nebula, it broke up into smaller nebulas, to each of which was imparted, by the impact, a small rotational motion. Through the force of its own gravity, one such smaller, spinning cloud began to contract. As it became smaller and smaller, by the law of conservation of angular momentum (the same one that makes a skater spin faster when the arms are held close to the body), the nebula's speed of rotation increased. This, in turn, caused it to flatten through centrifugal force. The action of gravity gradually caused most of the material to be concentrated at the center of the cloud. There, having reached a sufficiently high pressure and a temperature of about 6 million degrees Celsius, the nuclear reactions that led to the birth of our Sun were triggered.

The rest of the cloud, on the other hand, soon became a fairly thin disk (see figure at left). The dust in its inner regions, undergoing ever-increasing mutual collisions and subjected to the influence of magnetic and electrostatic forces, began to coalesce, forming clusters of granules about a meter wide. At this point, gravitational forces took over, and the aggregates attracted one another with increasing force. From these aggregates, larger bodies were formed, a few kilometers in size, known as planetesimals, orbiting around the Sun in variously inclined trajectories.

At the same time, the young Sun began to blow a strong wind of charged particles—protons and electrons—which almost completely stripped the inner part of the solar system of gases.

Afterward, further collisions among the planetesimals caused them to aggregate, forming bodies that were about the size of the Moon (3,000 km in diameter) and had roughly circular orbits. By slowly aggregating, these eventually formed the so-called "terrestrial" planets—Mercury, Venus, Earth and Mars. A Mars-sized planetoid that collided with our planet Earth later brought about the birth of our Moon.

Just beyond the orbit of Mars, however, the planetesimals were unable to combine to form a planet because of the strong gravitational perturbations caused by the large mass of Jupiter. The effects of the giant planet's gravitational influence were manifested in two ways: first, it prevented the aggregation of dust grains into larger lumps, hurling them instead in all directions, in toward the Sun and out to the edges of the so-

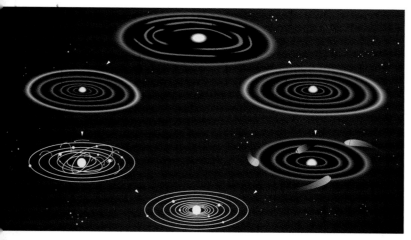

**Above, models of the two main hypotheses about the formation of the solar system. Both models begin with a rotating disk of dust and gas. The sequence on the left shows how planets might have formed through collisional aggregation. The right-hand sequence shows detachment of gas rings and the formation of gaseous protoplanets. The first hypothesis appears to work well for terrestrial planets and Uranus, Neptune and Pluto; the second, for Jupiter and Saturn.**

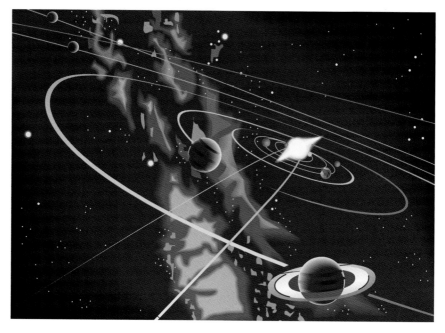

**Left, an imaginary representation of the solar system as it might appear to a hypothetical visitor from outside. The size of the planets has been exaggerated, because from such a distance, they would be visible only as tiny dots of light.**

lar system; then, it prevented any further aggregation of planetesimals that had already formed. The final result was that about 100,000 large pieces of rock remained between Mars and Jupiter—the asteroids, ranging in diameter from just a few meters to the 1,000-km diameter of the largest one, Ceres.

Far beyond Mars, things developed quite differently. Here, the influence of the solar wind was minor, so the gases of the protosolar nebula were not swept away. As a result of instability due to the gravitational effects of the Sun, rings of gas broke off from the edges of the nebula. These condensed gradually into large masses that were roughly spherical in shape—the gaseous protoplanets, from which the gas-giant planets Jupiter and Saturn were formed, with roughly the same chemical composition as the Sun (mainly hydrogen and helium).

The model of collisional aggregation described before, for the formation of the terrestrial planets, also works well for Uranus and Neptune. These two planets formed from planetesimals that were, however, not merely rocks but also contained ice, water, carbon dioxide, ammonia and methane that had condensed because of the great distance from the Sun and the consequent low temperatures. The protosolar nebula was less

dense in this region, and the embryos of Uranus and Neptune circled the Sun much more slowly than did Jupiter and Saturn. For these reasons, Uranus and Neptune were unable to capture as much primordial hydrogen and helium from the nebula as Jupiter and Saturn did. As a result, their size came to a halt somewhere between these two giants and the terrestrial planets.

As for the origin of Pluto and its moon Charon, a recent theory suggests that these two bodies came to be formed, like Uranus and Neptune, from the collisional aggregation of frozen planetesimal fragments. According to this theory, the same process of formation accounts for several dozen or even hundreds of tiny frozen mini-planets located beyond the orbit of Pluto. This hypothesis has recently been supported by research results, with the discovery of two dozen objects about 200 km in size in transneptunian and transplutonian orbits. According to Alan Stern at the University of Colorado, thousands of frozen mini-planets might have formed in the outer regions of the primordial solar system, objects about 1,000 km in diameter, which were then expelled from the system, leaving as sole survivors Pluto, Charon and Neptune's largest moon, Triton.

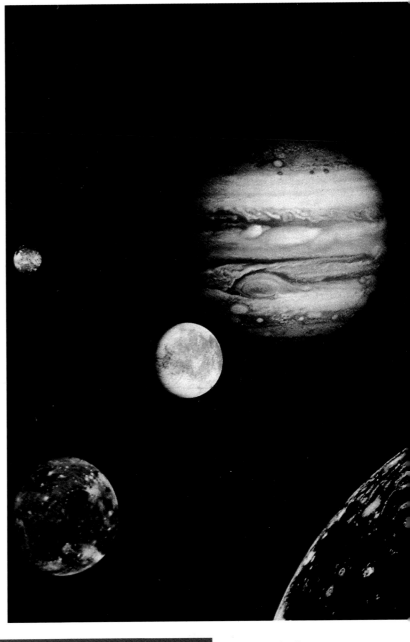

Above, Jupiter's system of moons looks like the solar system in miniature. The satellite systems of the giant planets probably formed in a similar manner to the solar system. This photomontage created by NASA shows the four largest moons of Jupiter: Io, Europa, Ganymede and Callisto.
Left, a table illustrating the planets' main characteristics.

## CHARACTERISTICS OF THE PLANETS IN THE SOLAR SYSTEM

| | Distance from the Sun (Earth = 1) | Equatorial diameter (km) | Mass (Earth = 1) | Period of rotation | Orbital period | Number of moons |
|---|---|---|---|---|---|---|
| Mercury | 0.3871 | 4 878 | 0.0553 | 58.65 days | 87.969 days | 0 |
| Venus | 0.7233 | 12 102 | 0.8149 | 243.01 days | 224.701 days | 0 |
| Earth | 1.0000 | 12 756 | 1.0000 | 23.9345 hours | 365.256 days | 1 |
| Mars | 1.5237 | 6 786 | 0.1074 | 24.6229 hours | 686.98 days | 2 |
| Jupiter | 5.2028 | 142 984 | 317.938 | 9.841 hours | 11.8623 years | 16 |
| Saturn | 9.5388 | 120 536 | 95.181 | 10.233 hours | 29.458 years | 20 |
| Uranus | 19.1914 | 51 118 | 14.531 | 17.9 hours | 84.01 years | 15 |
| Neptune | 30.0611 | 49 528 | 17.135 | 19.2 hours | 164.79 years | 8 |
| Pluto | 39.5294 | 2 320 | 0.0022 | 6.3872 days | 248.54 years | 1 |

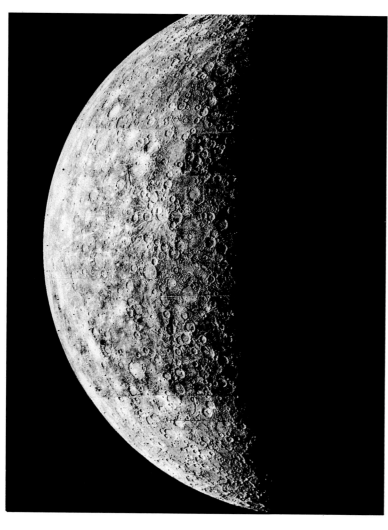

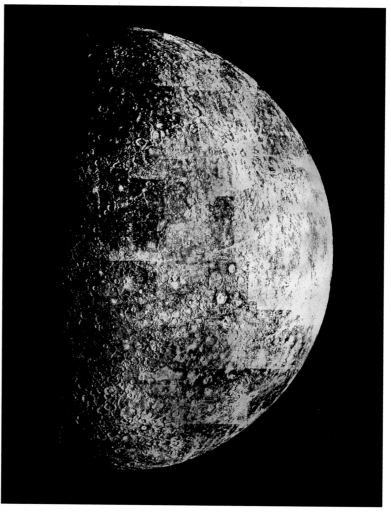

Opposite, four ultraviolet images showing the atmosphere of Venus, which is mainly composed of carbon dioxide. It is so thick as to appear impenetrable even to the cameras of interplanetary probes. In visible light, the surface appears uniform and devoid of features. Ultraviolet light, however, reveals the atmospheric dynamics and the evolution of clouds of sulfuric acid, which form curious shapes similar to a wedge, a Y and a reversed C.
■ Pictures taken in 1979 by the photopolarimeter on the Pioneer Venus probe.

Above and right, two hemispheres of Mercury taken in 1974 by the Mariner 10 probe, approaching and receding, respectively. Mercury appears similar to the Moon, with a large part of its surface covered with impact craters. The two hemispheres, however, reveal substantial differences: the one above has more craters, whereas the one at right shows vast areas that are smooth, and a large basin, 1,400 km in diameter—the Caloris Basin—is visible at the center of the photograph. This appears to be the only example of a feature resembling the large basins (known as "seas") on the Moon, which were formed by lava that flowed over the lunar crust 3 billion years ago. Because of the length of Mercury's day (equivalent to 2 of our months) and the absence of any atmosphere, Mercury has extreme temperature swings: 430 degrees C by day and -170 degrees C by night.
■ Each picture is a mosaic of 18 images taken from a distance of 200,000 km with the probe's wide-field (62 mm) camera. The picture on the right is a false-color enhanced image by Mehau Kulyk.

*Venus inaugurated the age of interplanetary missions in 1961 with the Soviet Sputnik 7 probe, which, however, failed. The following year, NASA had the first successful planetary flyby, sending Mariner 2 within 35,000 km of Venus. In 1966, the Soviet probe Venera 3 was the first space vehicle to touch down on another planet. In 1967, Venera 4 parachuted a module, which analyzed the atmosphere for the first time but which, once it had touched down, abruptly ceased to function. The same thing occurred with the next two probes, Veneras 5 and 6. In 1969, Venera 7 landed on Venus, transmitting data for 23 minutes and recording for the first time the environmental factors that had interfered with the previous probes: pressure of 90 atmospheres and temperature of 477 degrees C! In 1974, American probe Mariner 10 brushed past Venus on its way to Mercury and was the first to reveal details of the planet's atmospheric circulation. In 1975, Venera 9 and Venera 10 produced the first photographs of another planet's surface. In 1978, NASA launched the Pioneer Venus probe, whose orbiter accurately analyzed the atmosphere of the planet for 14 years. Unlike Venus, Mercury has been visited by only one probe, Mariner 10, which, using an ingenious "slingshot" technique, made three successive close approaches to the planet between March 1974 and March 1975.*

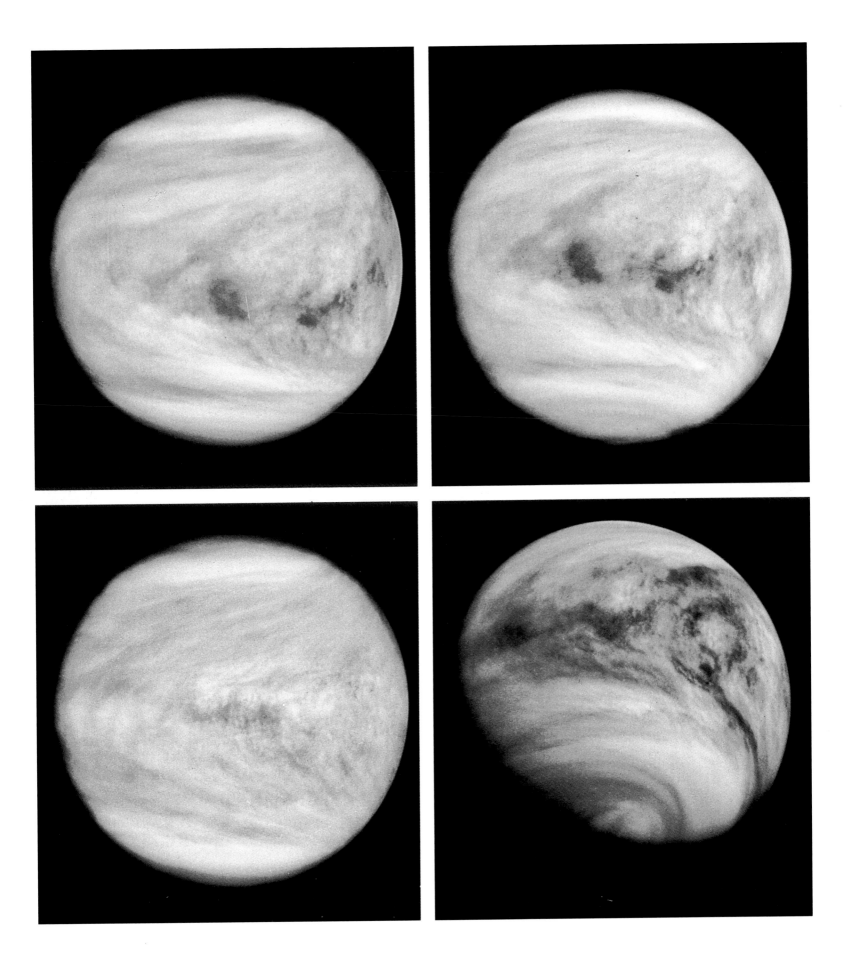

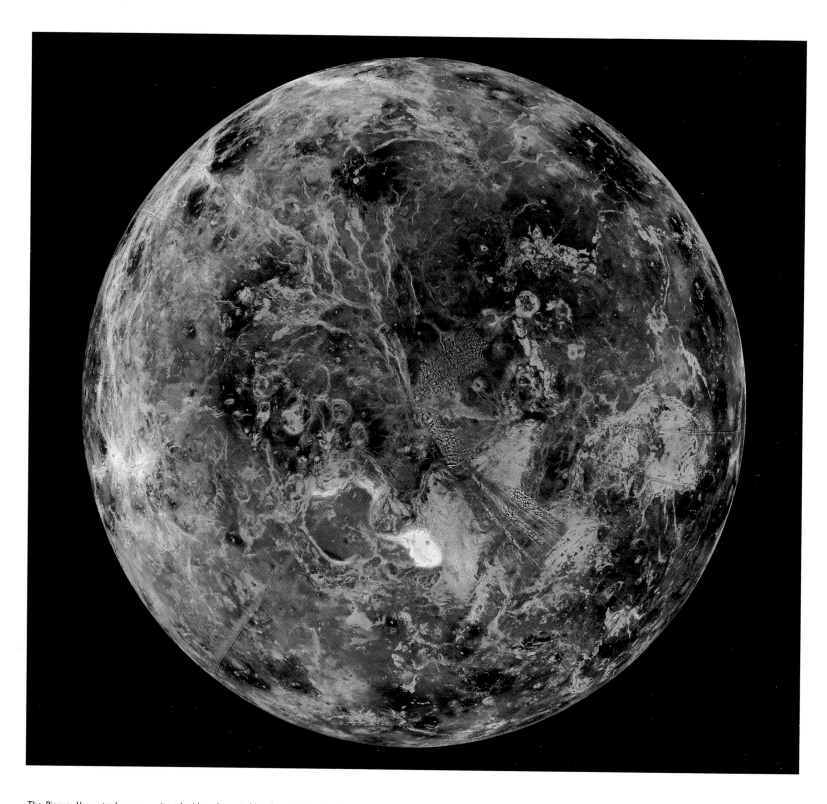

The Pioneer Venus probe was equipped with radar capable of achieving what conventional photographic techniques had never been able to do—to map the surface of Venus. The principle is very simple: Radio signals from the probe to the planet's surface bounce back from the ground at various angles, and the time delay between transmission and reception is measured. It is thus possible to reconstruct surface details. The first radar mapping of Venus had, however, been carried out from Earth between 1975 and 1977 by the 64-m radio telescope at Goldstone, California, and the 305-m dish at Arecibo, Puerto Rico, yielding a 4- to-20-km resolution. Although the Pioneer Venus probe's radar was unable to improve on this resolution, picking up details down to about 70 km in size, it did perform a more complete mapping of the planet. In 1983 and 1984, the Soviet probes Venera 15 and Venera 16 brought about a marked improvement in quality. They were fitted with a new kind of radar, known as SAR (synthetic aperture radar), which made it possible to improve horizontal resolution by one order of magnitude, picking up surface details 1 to 2 km wide.

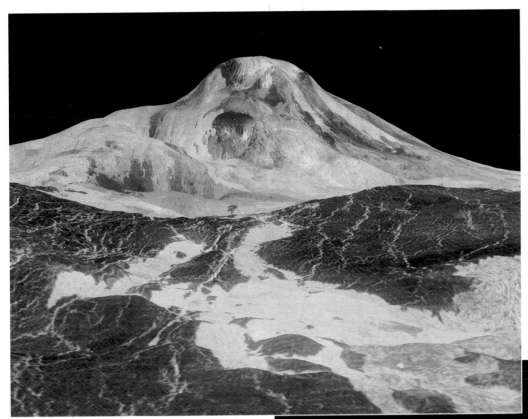

Left, the Maat Mons, a volcano of about 8,000 m in height, in a computer-simulated "3-D" image, as it would appear if seen from a height of 1,700 m. Flows of solidified lava spread over the base of the volcano for hundreds of kilometers. The presence of volcanoes and very extensive lava flows are the dominant surface characteristics of Venus. There is evidence, though not yet conclusive, that many of its volcanoes are active. ■ The images on this page were created using data from the Magellan probe's radar mapping of Venus, combined with a digital altimetric-radar map drawn up by the United States Geological Survey (USGS). The pictures were produced at the Multimission Image Processing Laboratory of JPL. Simulated colors have been used to reveal more of the fine surface details.

On facing page, a computer-generated mosaic of the northern hemisphere of Venus prepared by the Jet Propulsion Laboratory (JPL) in Pasadena, California, using radar images gathered by the Magellan probe. Between 1990 and 1992, Magellan completed three phases of radar mapping lasting 8 months each, during which 98 percent of the surface of Venus was mapped. The color is based on color photographs of Venus's surface produced by the Venera 13 and Venera 14 probes in 1982.

Right, a computer-simulated "3-D" view of the western part of an area of Venus known as Eistla Regio. The simulated vantage point is at a height of about 7,500 m. The two mountains that are visible are volcanoes: to the left, the Gula Mons, 3,000 m high; and to the right, the Sif Mons, 2,000 m high. The distance between the two is 730 km. Flows of fossilized lava stretch for hundreds of kilometers across the fractured plains visible at the base of the Gula Mons.

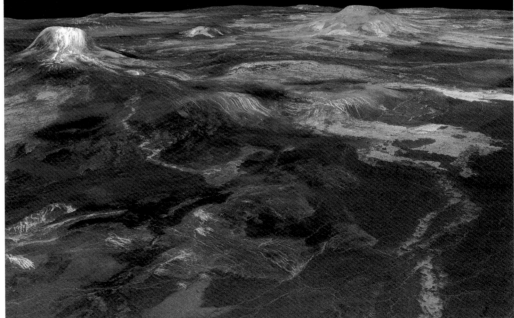

The final stage in the mapping of Venus was accomplished by NASA's Magellan probe, which began its work on September 15, 1990. This probe also was fitted with synthetic aperture radar and was able to improve the resolution of surface details by a further order of magnitude, taking the horizontal resolution to 120 m and vertical resolution to 10 m. On September 14, 1992, the third mapping cycle came to an end. The probe was then placed into a lower orbit for a further series of three cycles, during which the gravitational profile of the planet was measured to reveal anomalies in surface mass; at the same time, Magellan was also analyzing the density of the Venusian atmosphere. On October 11, 1994, with its mission completed, Magellan was crashed onto the planet's surface.

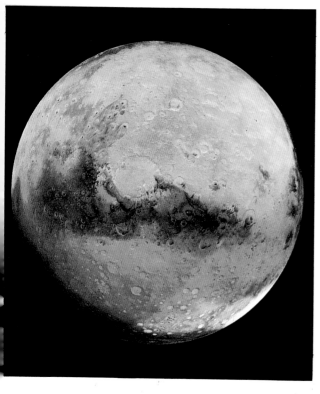

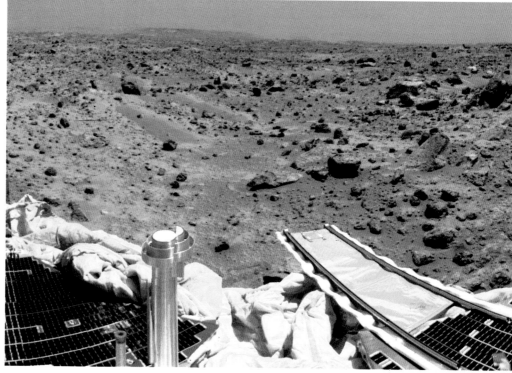

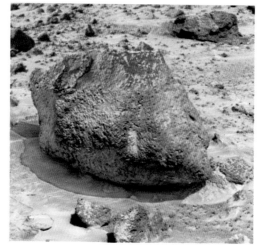

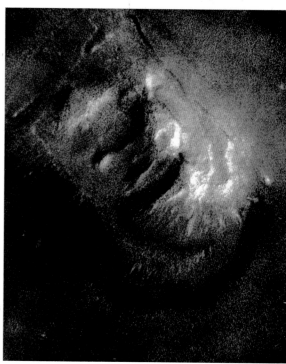

Unlike Venus, Mars has a very tenuous atmosphere composed of carbon dioxide, with a density a mere 1/150 of the Earth's atmosphere. The median ground temperature on Mars today is minus 50 degrees C, but the planet must have been far hotter in the past—so much so that water once ran freely on the surface. There are, in fact, traces of dried riverbeds and even of a huge, ancient northern-hemisphere ocean, with an estimated depth of roughly 1,500 m.

Above left, the area around Mars's Schiaparelli crater (near center of photograph), with the extended black band of Oxia Palus to the south. This composite image represents the view from 2,500 km above the Martian surface. The image was produced by Alfred McEwen, Tammy Becker and Lawrence Soderblom of the USGS from approximately 100 infrared and ultraviolet images taken with two 1,500-mm cameras aboard the Viking orbiters. Above right, a panorama of the Martian landscape surrounding the Mars Pathfinder probe that landed in 1997. Right, the boulder nicknamed Yogi, photographed by Pathfinder's camera. At far right, a partial view of the Cydonia sector reveals a new view of the famous "face" on Mars, photographed on April 6, 1998, by the Mars Global Surveyor orbiting probe. With the Surveyor's 4-m resolution, this feature proves to be a 2-km-wide grooved rocky plateau.

After two failed Soviet attempts—Mars 1 in 1962 and Zond 2 in 1964—the USA reached Mars in 1965 with Mariner 4, which sent 22 pictures back to Earth—the first close-up pictures of an alien world. This was followed in 1969 by Mariners 6 and 7, which took 75 and 126 photographs. In 1971, Mariner 9 functioned for 17 months in orbit around Mars, transmitting 7,300 photographs and a mass of data on the atmosphere, climate, geology and magnetic structure of the planet. The most fundamental work was carried out by the Viking 1 and Viking 2 probes. Each comprised an orbiter and a landing module, and both probes arrived at the planet in 1976. Vikings 1 and 2 functioned until 1980 and 1982 (despite a predicted life span of only 90 days), capturing 51,000 images and reams of data, analyzing the soil and mapping the Martian surface to a resolution of 50 m.

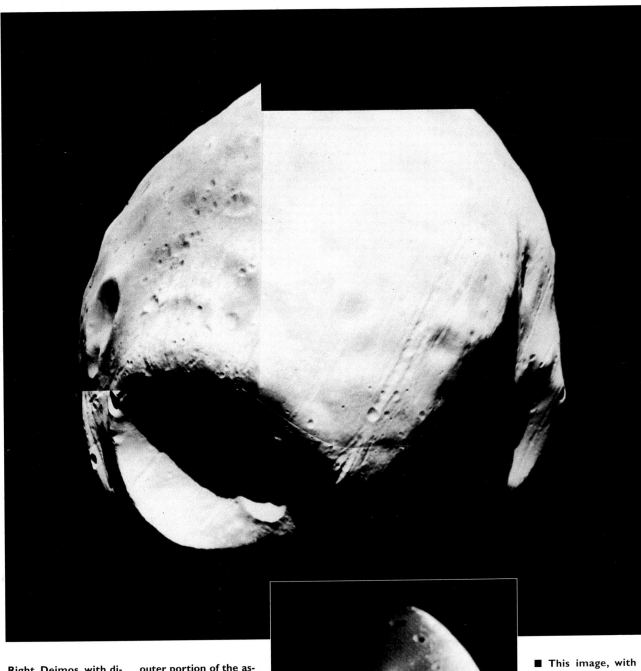

Mars has two tiny moons, Phobos and Deimos, which in all likelihood—given the planet's proximity to the asteroid belt—are two captured asteroids. This theory is supported by the resemblance between the Martian moons and the first asteroids to be photographed (see following pages).

Left, Phobos, which has the shape of an ellipsoid 28 km by 22 km by 18 km in size. The large crater visible at the bottom is called Stickney and has a diameter of almost 10 km. The lines that branch out from the crater have a width of 100 m to 200 m and a depth of 20 m. These are most probably surface cracks caused by the impact that formed Stickney.

■ Mosaic of images taken by Viking 1's orbiter at a distance of 612 km.

Right, Deimos, with dimensions of 15 km by 12 km by 11 km. Unlike Phobos, whose surface is fairly homogeneous, Deimos has bright spots. Phobos and Deimos are fairly dark, more so than is typical for asteroids in the inner portion of the asteroid belt, from which, in theory, they were captured. They are more like the tiny "Class C" planetoids of the outer portion of the asteroid belt. The composition of Phobos and Deimos seems to be similar to that of carbonaceous meteorites, in which the level of carbon is higher than in ordinary rocky meteorites and whose composition resembles that of the primordial solar-system disk.

■ This image, with a resolution of 200 m, is a combination of two shots taken in monochromatic light by the Viking 1 orbiter. Orange and violet filters were used to enhance the color of the moon, which is actually a dull gray, like that of Phobos.

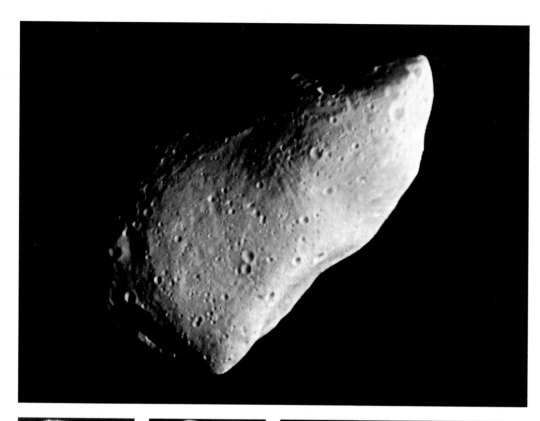

Left, the asteroid Gaspra photographed on October 29, 1991, by the Galileo probe. Gaspra is not a primordial body but originates from the fragmenting of a much larger asteroid. Its size is 19 km by 12 km by 11 km.

■ The picture was put together by imaging teams from JPL, the US-GS and Cornell University at Ithaca, using the black-and-white shot with the best resolution (54 m) taken at the Galileo probe's closest approach to Gaspra. Onto it were superimposed low-resolution (164 m) images taken in violet, green and near-infrared light to produce this false-color image, in which it is possible to pick out the surface topographical details.

Facing page, the asteroid Ida with its tiny moon Dactyl, photographed in visible light by the CCD camera on NASA's Galileo probe on August 28, 1993. Ida's dimensions are about 58 km by 23 km; Dactyl's diameter is about 1.5 km. Ida is part of a family of more than 200 asteroids known as the Koronis family, deriving from the fragmentation of a single, much larger body.

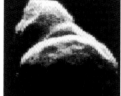

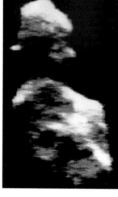

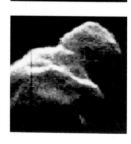

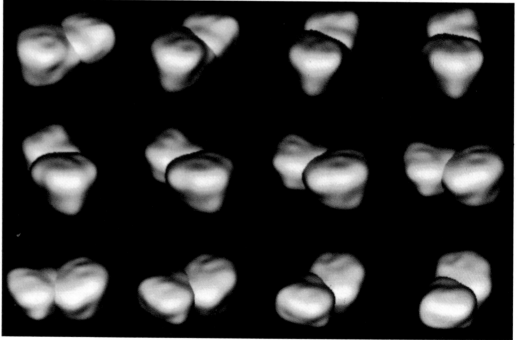

Above, three radar images of the asteroid Toutatis taken by Steven Ostro of JPL with the 70-m and 34-m radio telescopes at Goldstone, California, during the small asteroid's close approach to Earth (4 million km) in December 1992. Toutatis is made up of two bodies (4 km and 2.5 km in size) in contact.

Right, computer-enhanced radar images of the asteroid Castalia taken in 1989 during its closest approach (4 million km from Earth) by Steven Ostro, using the world's largest (305 m) radio telescope, at Arecibo. Like Toutatis, Castalia is made up of two bodies in contact but, in this case, smaller ones measuring just 800 m to 900 m each.

In 1771, French mathematician Joseph-Jérome de Lalande—followed in 1772 by Johann Daniel Tietz, professor of mathematics at the University of Wittenberg—noted an empirical relationship in the distances of the planets from the Sun. Also in 1772, the very young Johann Elert Bode (later director of the Astronomical Observatory of Berlin) showed that one term was missing from this relationship, corresponding to a planet that, according to his calculations, should be located between Mars and Jupiter. When William Herschel discovered Uranus in 1781, it was noted that this planet too followed the relationship, now known as Bode's law, or the Titius-Bode law (Titius was a Latinized form of Tietz). In the late summer of 1800, Austrian baron Franz Xaver von Zach, an expert amateur astronomer, formed a research team composed of some of the most famous astronomers of the day, including Bode himself, as well as Schröter, Olbers and Harding, to try to find the missing

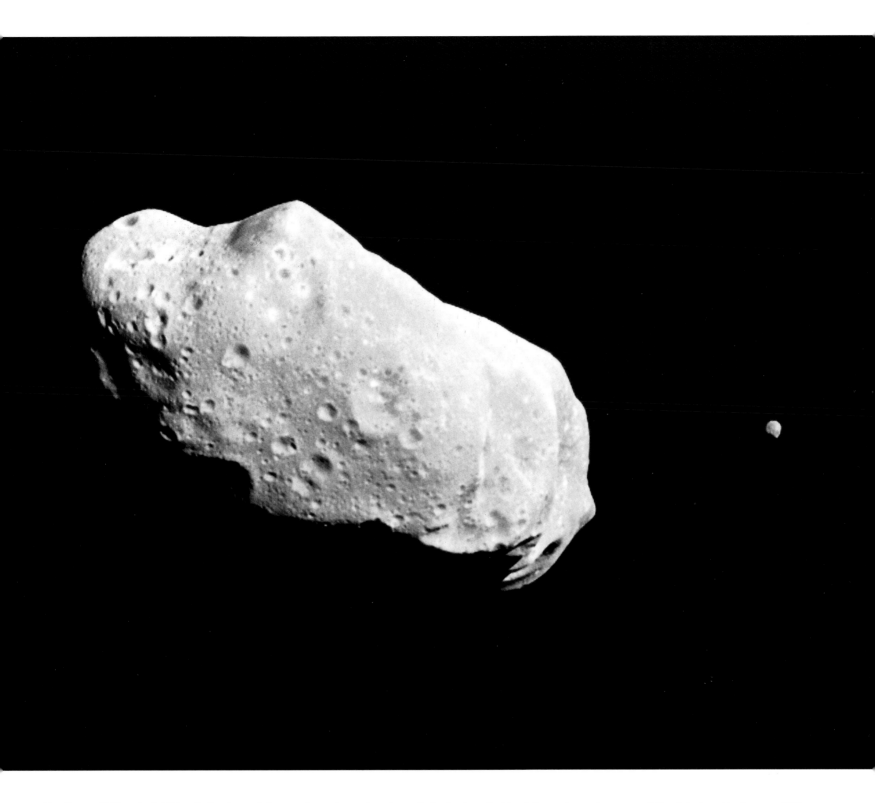

object. However, this *Himmel-polizei* ("sky police"), as it was called, was preceded by Italian astronomer Giuseppe Piazzi. Director of the Astronomical Observatory of Palermo from 1787, Piazzi had been engaged for 9 years in research into the proper motions of stars, a program suggested to him by Herschel, whom Piazzi had met in England when he made the chance discovery of a new object, the asteroid Ceres, on January 1, 1801. Before it was possible to calculate its orbit, Ceres had un-

fortunately moved too close to the Sun for observation. Nevertheless, using the few observations that had been carried out by Piazzi, the great German mathematician Carl Friedrich Gauss was able to calculate Ceres' orbit and to predict its successive positions in the sky so that, on January 1, 1802, exactly one year after its discovery, Baron von Zach was able to find it.

It soon became obvious that Ceres could not be the "missing planet,"

because it was too small (Herschel estimated its diameter at 250 km, which was actually four times smaller than the asteroid's true size). Meanwhile, other planetoids were soon found. It was Olbers who discovered the second asteroid, Pallas, on March 28, 1804. Harding discovered the third one, Juno, in 1804, and in 1807, Olbers found the fourth one, Vesta. To date, the orbital parameters of about 6,000 asteroids are known.

The atmosphere of Jupiter, fairly turbulent and brightly colored, is divided into light zones and dark belts. The lighter zones are formed by high clouds of icy ammonia crystals. The dark belts generally indicate lower-lying atmospheric regions made up of hydrogen and compounds like ammonium hydrosulfide.

■ This picture was taken by Voyager 1 at a distance of 30 million km.

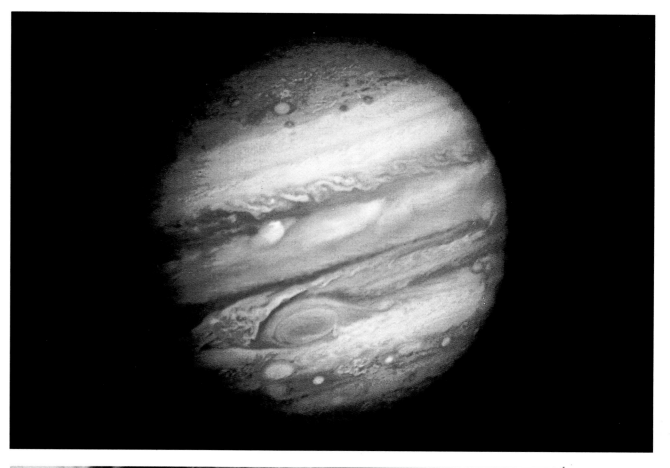

The Great Red Spot. Like the smaller dark red spots that occasionally become visible in Jupiter's atmosphere, the Great Red Spot is a hurricane-like weather formation, with an anticyclonic direction of circulation. The Great Red Spot is somewhat higher than the surrounding atmosphere; its color comes from red phosphorus, a product of the decomposition of phosphine.

■ A false-color enhancement of an image taken by Voyager 1 from 1.8 million km.

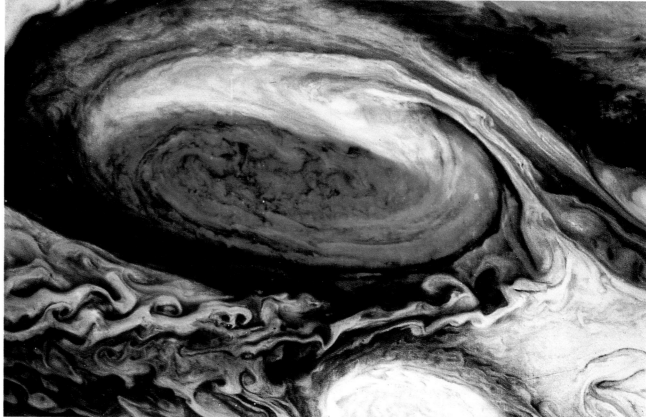

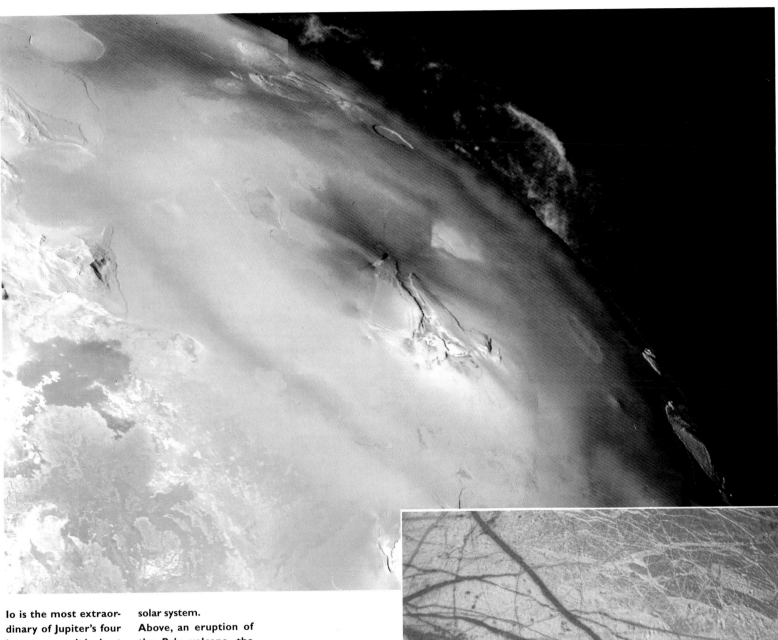

Io is the most extraordinary of Jupiter's four largest moons. It is about the same size as Earth's Moon, and its distance from Jupiter is just slightly more than that of the Moon from Earth. The satellite's proximity to Jupiter unleashes fearful tidal flexing effects in Io, causing an intense heating of its interior and creating an impressive amount of volcanic activity. There are about 10 volcanoes active on Io at any one time, making it the most geologically active body in our solar system.

Above, an eruption of the Pele volcano, the first to be discovered, showing a plume similar to that of a geyser. The sulfur dioxide prominence reaches a height of 300 km, and the material belched out has formed a structure 1,400 km in diameter.

■ A composite image created by Alfred McEwen of the USGS by combining high-resolution black-and-white photographs and low-resolution color photographs taken by Voyager 1.

Above, Jupiter's moon Europa displays an intricate web of cracks produced when the enormous ocean that originally covered the moon froze over and the surface fractured. The cracks were subsequently filled by ice mixed with dark silicas extruded from the interior.
False-color mosaic produced from infrared images taken by the Galileo probe on November 6, 1996.

Right, Saturn with six of its moons: Dione, in the foreground; Enceladus and Rhea to the left of Saturn; Tethys and Mimas to the right; and in the background, Titan. These moons have practically circular orbits that scarcely tilt from the ecliptic plane; they were formed from the disk of dust and gas that was present around the planet Saturn during its formation. The other 14 moons of Saturn, including three somewhat conspicuous ones—Iapetus, Phoebe and Hyperion—were probably captured by Saturn at a later stage. All are made up of ice, ammonia, silicas and methane compounds.
■ NASA photomontage using images from Voyager 1 taken in November 1980.

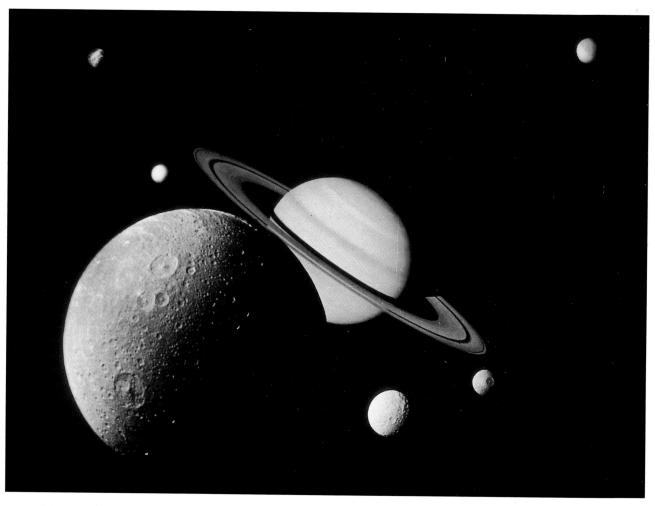

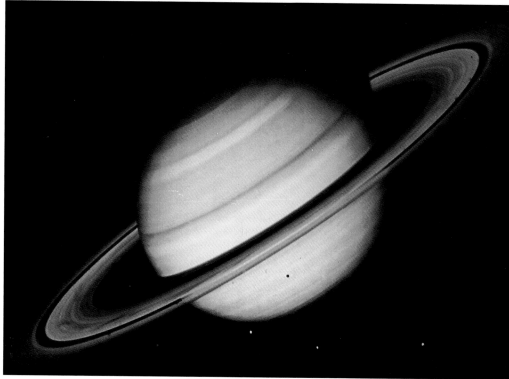

Saturn is famous for its system of rings, a distinction that it shares to some extent with the other gas-giant planets (although the rings of Jupiter, Uranus and Neptune are very meager in comparison). Saturn's ring system is 270,000 km in diameter but less than 1 km thick. The planet's atmosphere is divided into light zones and dark belts like those of Jupiter, but its appearance is much more uniform and less colorful, probably due to certain differences in the structure of its cloud layers, as compared with Jupiter's. The atmospheric composition is similar to that of Jupiter, consisting mainly of hydrogen and its compounds (such as methane, water and ammonia) and helium. Curiously, Saturn is not very dense—only 0.7 gram per cubic centimeter, which means that if there were an ocean large enough, Saturn could float on it.
■ This picture was taken by Voyager 2 at a distance of 18 million km from Saturn on April 8, 1981.

Four rings, at best, are visible through terrestrial telescopes, but the Voyager probes have revealed that each of the main ring components is made up of several subunits (as can be clearly seen in the picture at right), which become more and more evident the closer one gets to the planet. At the Voyager probes' closest approach, 10,000 rings were observed, arranged like the tracks of a record disk. The rings are made up of billions of chunks of rock and ice, varying from just a few microns in size to a few kilometers, each in its own orbit around Saturn.

■ False-color-enhanced image based on shots taken by Voyager 2 in visible, ultraviolet and orange light, at a distance of 8.9 million km.

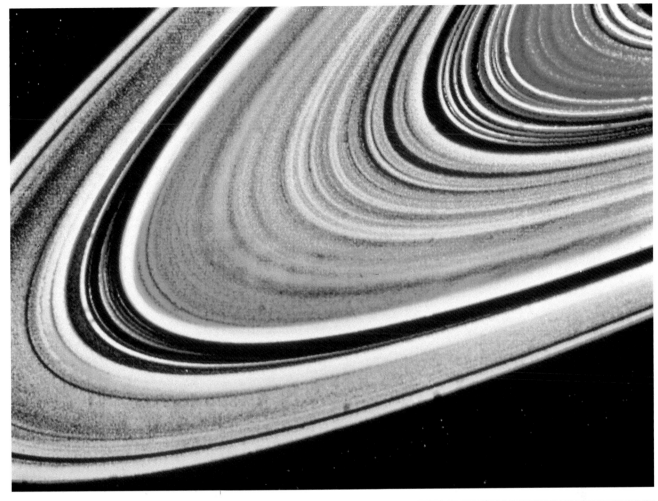

Right, the Wide Field/ Planetary Camera 2 on the Hubble Space Telescope is able to penetrate the thick nitrogen atmosphere of Saturn's largest moon, Titan, which eluded even Voyager 1's cameras.

■ Composites of images taken in October 1994 by Peter Smith, of the University of Arizona, in near-infrared, in which Titan's atmosphere is more transparent.

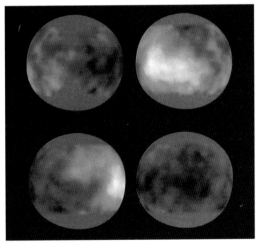

Right, Saturn's moon Enceladus, which, though 10 times smaller than Titan (500 km in diameter as opposed to 5,000 km), has an incredible variety of surface features, evidence of intense geologic activity in the past.

■ False-color mosaic of Voyager 2 images, taken at a distance of 119,000 km.

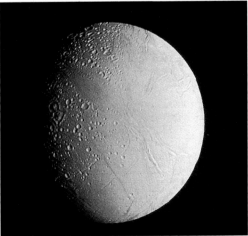

The first space vehicles to venture toward the outer solar system, beyond the asteroid belt, were the twin US probes Pioneers 10 and 11. Launched in 1972 and 1973, respectively, the first reached Jupiter at the end of 1973, and the second, having flown by Jupiter in 1974, reached Saturn in 1979. Their task, besides carrying out the first close reconnaissance of the giant planets, was above all to pave the way for two subsequent, more sophisticated probes and to prove that crossing the asteroid belt was much less risky than one might have imagined. For their more ambitious missions, Voyagers 1 and 2 were equipped with more sophisticated instruments (including 200-mm and 1,500-mm video cameras), and the Voyager probes have left an indelible mark on the history of space travel. Launched in 1977, Voyager 1 flew close by Jupiter in March 1979 and reached Saturn in November 1980. Voyager 2 kept its appointments with the two planets in July 1979 and August 1981. The exploration of the two giants of the solar system continues with two orbital probes. Galileo reached Jupiter in December 1995, while Cassini, now under way, is scheduled to reach Saturn in June 2004.

Right, two images of Uranus: one in true color and one in false color, centered on the planet's south pole. The planet's atmosphere is very opaque and apparently devoid of features because of the absorption of red light by the methane present in the upper layers, which gives the atmosphere its characteristic blue-gray color. False-color processing is needed to bring out a few details: a vast, dark polar cap becomes visible, surrounded by concentric bands, somewhat reminiscent of the atmospheric circulation patterns seen on Jupiter and Saturn.

■ The left-hand image is a composite of shots using blue, green and orange filters. For the right-hand image, ultraviolet, violet and orange filters were used. (Taken by the 1,500-mm camera on Voyager 2 from a distance of 9.1 million km.)

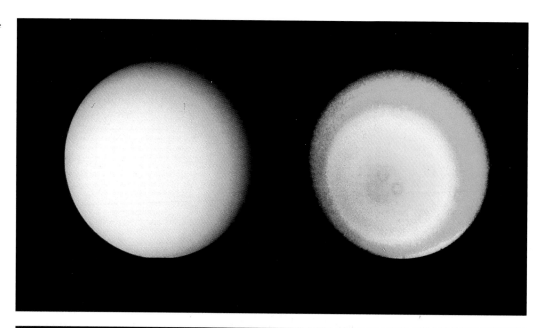

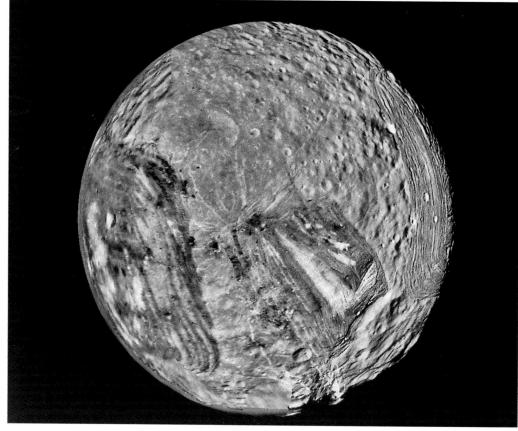

Below left, Uranus's moon Miranda. Uranus has a system of satellites that is quite similar to Jupiter's and Saturn's. The five largest moons of Uranus are Miranda, Ariel, Umbriel, Titania and Oberon, in order of their distances from the planet. They range from 480 km to 1,594 km in diameter. All show unexpected signs of past geologic activity, perhaps due to temperatures lower than the freezing point of water, which created their frozen, cracked surfaces. They also show the effects of subsequent gentle warming due to radioactive isotopes in their interiors, the tidal forces exerted by Uranus and the kinetic energy of impacting asteroids. The most interesting moon of Uranus is the smallest one, Miranda (left), which reveals a hodgepodge of surface features, including craters, valleys, mountains, escarpments and fractures, suggesting that this is a body that must at some time have broken apart due to repeated impacts, after which the fragments subsequently joined together again through mutual gravitational attraction.

■ Mosaic of nine pictures taken by Voyager 2 at close range.

*After Voyager 1's photographs of Saturn's moon Titan in November 1980 showed that a thick atmosphere concealed Titan's surface, NASA felt it would be useless to send Voyager 2 to do the same thing again. Therefore, instead of having Voyager 2 do a second reconnaissance of Titan, NASA revived an idea that had first been suggested in the 1960s. It involved making Voyager 2 continue on toward Uranus and Neptune and appeared*

*to stand only a 30 percent chance of success. The vehicle had not been designed to last for so long and to operate at such a distance from Earth, in total autonomy and in a much more distant region of the solar system than that of Jupiter and Saturn, a region where the light from the Sun would be considerably weaker. The probe had also suffered a number of problems, including failures of its radio antenna and its instrument platform.*

*All of this notwithstanding, Voyager 2 succeeded in completing the most spectacular solitary journey in the history of astronomy. In what became known as the planetary Grand Tour, Voyager 2 reached Uranus in January 1986 and Neptune in August 1989, giving us wonderful pictures of those faraway worlds.*

Right, Neptune's atmosphere appears blue because of its red-light-absorbing methane, which is present here in even greater quantity than on Uranus. The atmosphere of Neptune has essentially the same composition as the other gaseous planets, mainly hydrogen and helium. The Great Dark Spot visible in this photograph is a cyclonic feature similar to Jupiter's Great Red Spot and about the size of the Earth.
■ Composite of two images taken by Voyager 2.

Triton is the largest of Neptune's eight known moons. Below, Triton reveals a tortured appearance caused by the surface disruptions that resulted from the energy released as a result of its capture into orbit around Neptune. The consequent heat generated in the moon's interior undoubtedly fueled many vast volcanoes on Triton. Evidence of these can still be found in the persistence of two geyser-like features that have been identified in the area around Triton's south pole.

■ Mosaic of 18 high-resolution black-and-white images and 6 low-resolution color images taken by Voyager 2.

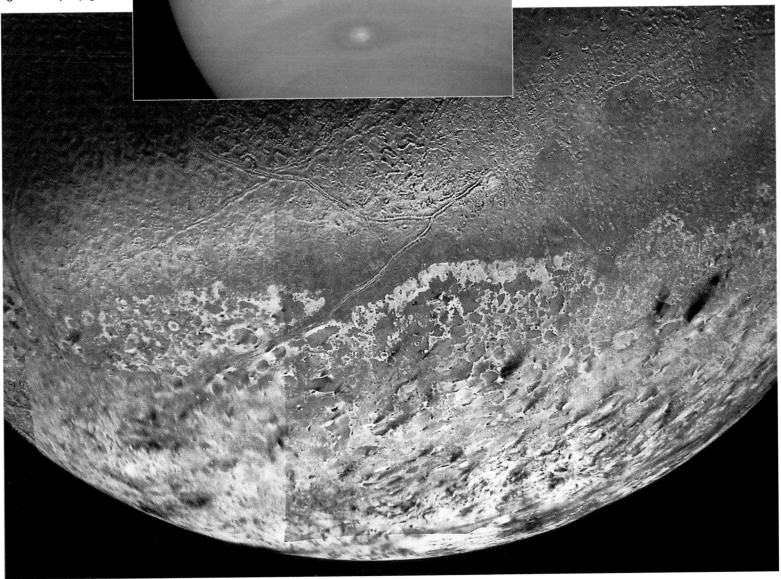

Right, Pluto and its moon Charon photographed on February 21, 1994, by the Hubble Space Telescope. This is the most detailed picture ever taken of this far-away system, the only one in which the two bodies are seen clearly as two definite disks. Note Pluto's lighter color: its crust is made up of light blue frozen methane, whereas Charon's is formed from frozen water covered by dark carbon deposits.

■ Photographed in ultraviolet light by Rudolf Albrecht (Space Telescope European Coordinating Facility) using the Hubble's Faint Object Camera. The picture was taken when the separation between Pluto and its satellite was 0.92 arc-second.

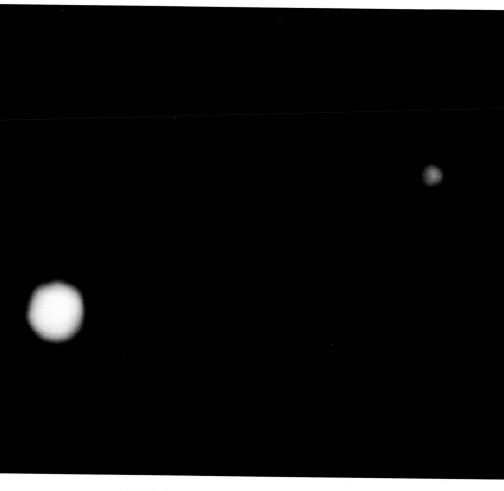

On facing page, an artist's rendering of the Kuiper belt, a region of the solar system beyond the orbits of Neptune and Pluto. The region is said to be filled with billions of cometary objects varying in size from a few kilometers up to a few hundred kilometers. The belt's existence was predicted in the 1950s by the Dutch astronomer Gerard Kuiper. It now appears to be confirmed by the recent discovery of the very distant objects discussed on pages 19 and 35.

This is the first map of the surface of Pluto. It was created by Alan Stern of the Southwest Research Institute and Marc Buie of Lowell Observatory using a computer to combine and enhance four images taken by the Hubble's Faint Object Camera in June and July 1994. The map, which covers approximately 85 percent of the surface, confirms that Pluto has a dark equatorial band and brilliant polar caps, just as observations during the mutual eclipses of Pluto and its moon Charon in the 1980s had indicated. The variations in brightness may be due to surface structures, such as basins or impact craters, or to transitory seasonal changes.

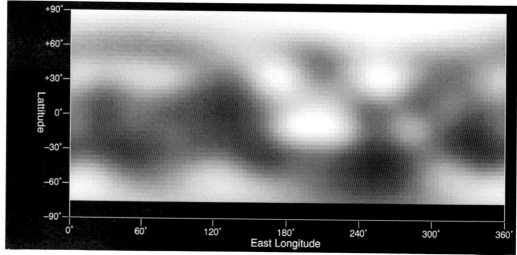

*Pluto was discovered in 1930, following extensive observations of the sky carried out to locate an object that was believed to be responsible for apparent perturbations in the orbits of Neptune and Uranus. Subsequently, it was discovered that the apparent perturbations were not real. Pluto is, after all, too small to produce any significant gravitational disturbances of the other planets. After its discovery, the distant planet was all but forgotten, neglected even in projects that explored nearby regions. As irony would have it, however, on July 6, 1978—10 months after Voyager 2 was launched—James Christy announced the discovery of Charon. In 1985, a 5-year cycle of eclipses and occultations between Pluto and its moon began. Observations of these events enabled the dimensions and even, to some extent, the surface features of Pluto and Charon to be determined, revealing that the two constitute a unique case of a "double planet" within the solar system. Recently, it has also been discovered that the increased heat Pluto received from the Sun as it reached perihelion in 1989 is causing part of Pluto's frozen methane crust to melt, creating a fairly extensive atmosphere. There is more than enough evidence to warrant sending a probe to the planet without delay. Many such projects are in the works, and it seems likely that a probe will be launched in the early years of the 21st century to reach Pluto within about 10 years.*

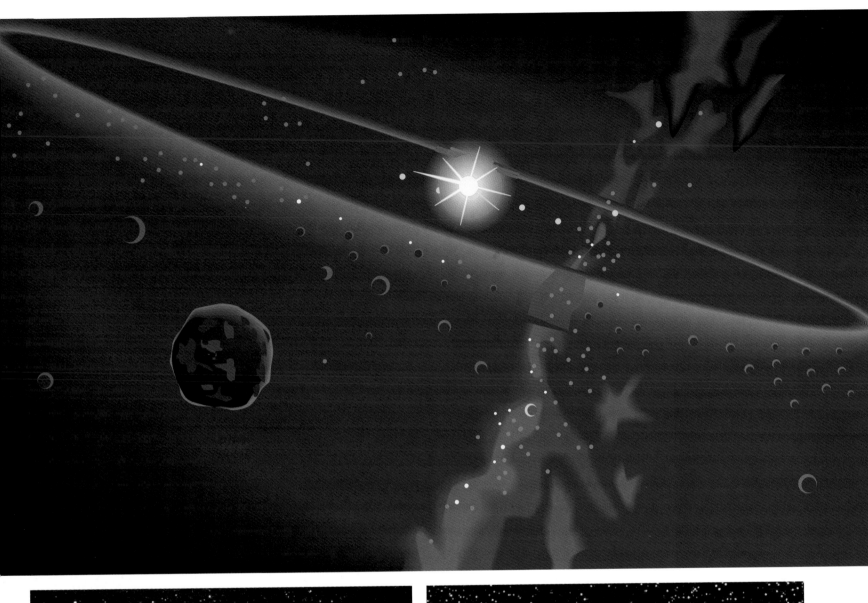

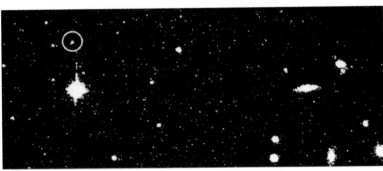 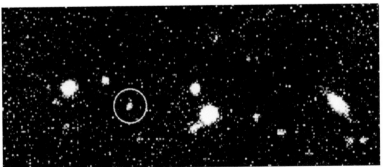

**Above, the first two transplutonian asteroids (or comets), 1992 QBI (left) and 1993 FW (right), discovered in September 1992 and in March 1993, respective-** **ly, photographed at the ESA (with the NTT and the 1.5-m Danish Telescope, respectively). They have been named Smiley and Karla and are about 200 km in size.**

In 1951, Gerard P. Kuiper put forward the hypothesis that at an early stage in the formation of the solar system, a belt of comets about 2 billion km wide formed at a distance from the Sun around the orbit of Pluto. According to this theory, most of these comets then aggregated, forming the planets Uranus and Neptune. The fragments that were not involved in forming the planets came to inhabit a vast area beginning just beyond the orbit of Neptune. Perturbations caused by the giant planets' gravity very gradually alter their orbits, drawing them back toward the inner regions of the solar system. This process accounts for the "short-period" comets, comets with orbital periods of less than 200 years.

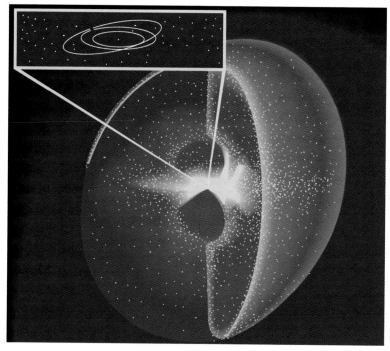

Right, artist's rendering of the Oort cloud, a region at the far outer fringes of the solar system that contains billions of comets. At this great distance, the comets are not affected by the gravitational action of the planets. Since the Oort cloud stretches almost halfway to the nearest stars, however, its comets are subject to the gravitational influences of those stars. The middle diagram shows how the close approach of a nearby star can tear comets away from the solar system. The diagram at far right shows an alternative scenario,

in which the gravity of a passing star causes a comet to plunge inward toward the orbit of our own planet Earth. When this happens, the comet begins a multi-million-year journey toward the Sun. If it comes close

enough, the Sun's warmth can sublimate the ices that make up the comet's nucleus, producing a large coma and (with some help from the solar wind) a tail of enormous length.

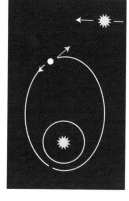 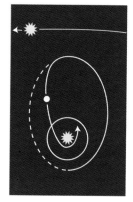

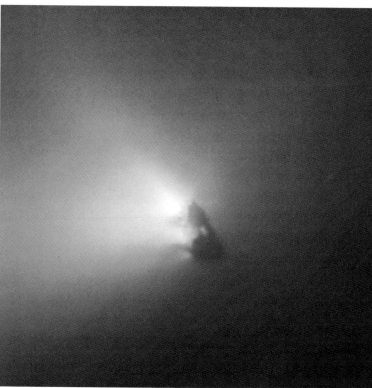

Left, the nucleus of Halley's comet, the only comet to date to have been explored at close range, in a photograph taken in March 1986 by the ESA's probe Giotto, which "brushed past" Halley at a distance of 600 km. The nucleus is shaped like a large potato, 16 km by 8 km by 8 km in size, and its surface is highly irregular and pockmarked by holes. Clearly visible are spots where the heat of the Sun is violently sublimating the ices that cover the nucleus, producing spurting fountains that create a halo of gas and dust around the comet. The density of the comet is extremely low, about 0.2

gram per cubic centimeter, five times less than the density of water. Its composition includes, besides frozen water, carbon dioxide, methane and ammonia, silicaceous dusts and organic compounds. It is the latter, carbon-based compounds that cover much of the nucleus with a dark crust that makes Halley's comet the darkest known body in the solar system. Data from the Giotto probe confirm the notion, suggested in 1950 by American astronomer Fred Whipple, that a comet is a kind of "dirty snowball," an aggregate of frozen matter mixed with primordial dust from the peri-

od when the solar system was formed. This model was complemented in 1986 by another model, which was put forward by Paul R. Weissman of the Jet Propulsion Laboratory and referred to as "a heap of primordial debris," whereby a comet nucleus is said to be an amalgam of small fragments weakly held together by ice. ■ A composite of six images taken with Giotto's Halley Multicolor Camera.

*Following the formation of Uranus and Neptune, the remaining cometary fragments, having more or less circular orbits, arranged themselves within the Kuiper belt. Long-term perturbations from the forming embryos of Uranus and Neptune gradually increased both the eccentricity and the size of the comets' orbits. Then, perturbations caused by the combined gravity of the stars, gas and dust in our entire galaxy began to make themselves felt. The result was that the comets were gradually pulled farther away from the Sun, up to average distances of 1.5 trillion km. Thus was formed the inner region of the Oort cloud, an enormous shell containing billions of comets, whose existence was first hypothesized in 1950 by Dutch astronomer Jan Oort. From there, continued perturbations by galactic matter pushed many comets to distances of 7 trillion to 20 trillion km from the Sun, into the outer region of the cloud. According to the most recent estimates, the Oort cloud is an enormous spheroid with dimensions of 30 trillion by 24 trillion km (3.2 by 2.5 light-years). The outer region of the Oort cloud is said to contain 100 billion to 1 trillion comets; and the inner region, 500 billion to 1 trillion comets.*

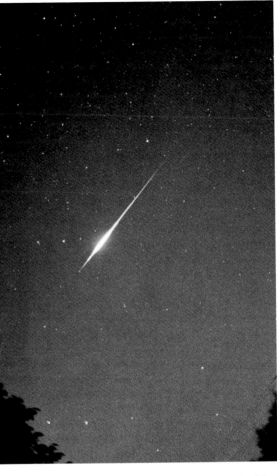

Left, a "fireball," or bolide (extremely bright meteor), that appeared on the night of August 11/12, 1993, during the Perseid meteor shower. The meteor reached a brightness comparable to that of the full Moon. ■ Photographed by Severino Lodi, from Soliera (near Modena), Italy, using a 24-mm f2.8 lens.

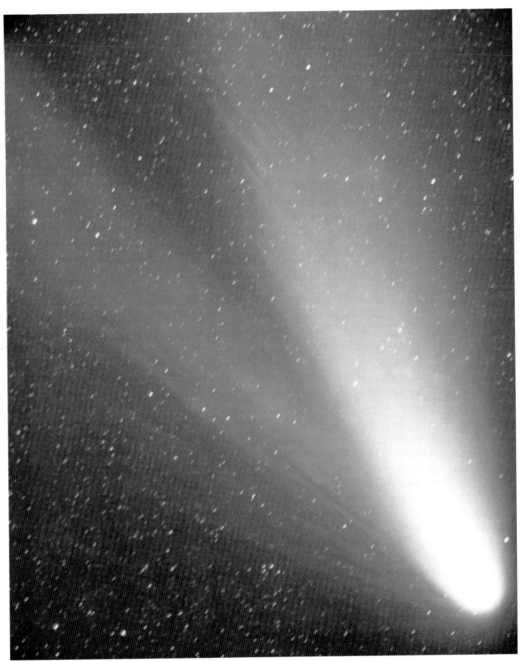

When a comet comes fairly close to the Sun, the action of the solar wind (a flux of charged particles that is the residue of the stormy primordial wind referred to on page 18) and the pressure of sunlight combine to produce the comet's tail. Sublimation of the frozen matter in the comet's nucleus produces the gases that form the "gas tail." A second, distinct tail, called the "dust tail," is composed of dust particles shed by the comet. The dust tail is usually the more conspicuous of the two, reaching a length of tens or even hundreds of millions of kilometers. The 15-minute exposure at right shows the double tail of Comet Hale-Bopp.
■ Photograph taken by Eraldo Guidolin from the Dolomite Alps in Italy on March 16, 1997, with a 760-mm Flat Field Camera at f4, using Ektacolor Gold 1000 film.

The dust shed by periodic comets during their innumerable journeys around the Sun settles along their orbits. When the Earth crosses a comet's orbit, it passes through clumps of this dust. The dust particles burn up from friction with the Earth's upper atmosphere, at a height of 80 to 120 km. As a speck of this dust (somewhere between the size of a large grain of sand and a small soup bowl) evaporates, the gases produced by its evaporation ionize the atmospheric gases, creating a luminous trail—a "shooting star," or meteor. The Earth encounters a certain amount of interplanetary dust every day, giving rise to what are known as sporadic meteors. But when the Earth passes through a dense clump of dust such as that left behind by a comet, meteor showers occur, in which a few hundred meteors an hour can be seen.

# The stars.

The idea that the Sun is a star and that the myriad stars we see in the sky are also suns appears rather obvious to us today, but this was not always the case. In ancient times, it was believed that the stars were holes in a celestial sphere, through which it was possible to see the great fire that blazed outside. That is a concept which we would now dismiss as naive but which undoubtedly seemed appealing and rather ingenious at the time. The ancient Greeks and Romans defined anything that could be seen in the sky as "stars." However, they did make a distinction between stars which appeared to move—the planets (in Greek, *planetes* means vagabond, thus "wandering stars")—and "fixed stars," which maintained stable positions with respect to one another. Then there were "stars with hair"—the comets (from the Greek *kome*, meaning "hair")—and "new" stars, or novas, which seemed to appear suddenly. Chinese astronomers used even more intriguing terms for these last two categories: novas were called "guest stars," while comets were "sweeping (broom) stars."

The notion that the Sun and the stars were identical types of things was suggested by the philosophers of antiquity and revived during the Middle Ages by Nicola Cusano and in the Renaissance by Giordano Bruno. It gained impetus with the acceptance of the Copernican system: the Sun was no longer a "planet" like the others but became the fulcrum of the solar system, the pivot of planetary motion, the light that illuminated the world. Inevitably, stars thus came to be thought of as other suns too distant to be resolved into disks.

Scientists have for a long time wondered how the Sun manages to produce all the energy that we observe. Even after the great Italian astronomer Gian Domenico Cassini succeeded in calculating its distance in 1672 and its enormous size finally came to be understood (its volume was estimated at more than a million times that of the Earth), it was clear that if the Sun were composed of ordinary combustible matter, it would have burnt itself out within a few million years. This may have been in keeping with the scant geological knowledge of the period, whereby the Earth was reckoned to be of no great age. But during the 19th century, when it became clear that our planet was at least a few hundred million years old, such reckonings no longer made sense. When physicists Lord Kelvin and Hermann von Helmholtz suggested that solar energy was produced by a slow gravitational contraction of the Sun—a process that could have fueled the stellar furnace for no more than about 15 million years, according to von Helmholtz's calculations—this, too, failed to work, in the light of the newer knowledge about the great age of the Earth.

Only in 1927 did the correct route to the

**Below, a diagram showing the proton-proton chain reaction, which produces the energy that fuels the Sun's nuclear furnace and those of most of the stars for the majority of their life cycle. The blue circles are protons; the brown ones are neutrons.**

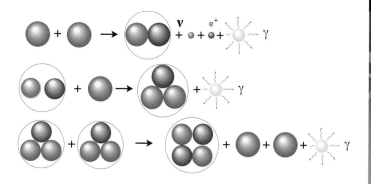

**Right, a diagram of the Sun in cross-section. Proceeding from the interior to the surface, one can see the core, the radiation zone, the convection zone and the photosphere.**

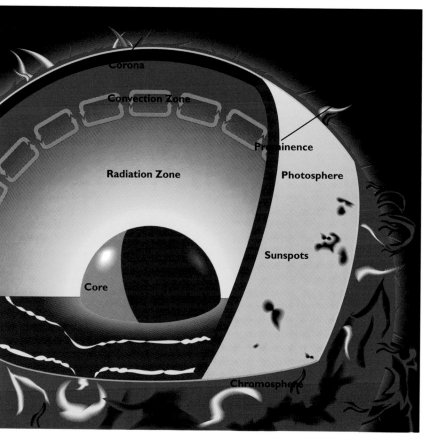

answer finally appear, when R. d'Escourt Atkinsons and F.G. Houtermans reasoned that the energy of both the Sun and the stars was produced by nuclear fusion reactions of light atoms. Not long after this, Carl Friedrich von Weizsacker speculated on the role of hydrogen nuclei, and in 1938, Hans A. Bethe and C.L. Critchfield developed the complete theory of the proton-proton chain. (A proton is an elementary particle with a positive charge, the only component of the nucleus of an atom of hydrogen, which contains no neutrons.) In the proton-proton chain reaction, two hydrogen nuclei join to form one deuterium nucleus (consisting of one proton and one neutron); the deuterium nucleus combines with another hydrogen nucleus, forming a nucleus of helium-3 (with one neutron and two protons); finally, two nuclei of helium-3 combine to make up a nucleus of ordinary helium (helium-4, with two neutrons and two protons), plus two hydrogen nuclei (which are then available to combine together to fuel further proton-proton reactions).

Other products of the proton-proton chain reaction are: neutrinos, elementary particles that are electrically neutral and have almost zero mass (indicated by $\nu$ in the left-hand figure on page 38); positrons, anti-electrons with a positive charge (indicated by e+); and gamma rays,

the radiation with the highest energy content (indicated by $\gamma$).

These reactions can occur at a temperature of 6 million degrees C. Such temperatures are reached within the core of the Sun, which, according to the latest theories, is at 14 million degrees C. From the Sun's core, the gamma rays travel slowly toward the surface. Colliding and interacting with millions of particles along the way, they weaken and are gradually transformed into X-rays, ultraviolet rays, visible light and infrared light. Most of the observable energy emanating from the Sun's surface is in the form of visible light. The Sun's core is surrounded by an area called the radiation zone, where energy transmission is by radiative processes; that is, each particle of matter absorbs light and re-emits it at a higher wavelength (see right-hand figure on page 38). Still farther out is the convection zone, where energy is carried by convective motion, similar to that which makes water boil in a saucepan. At the border between the radiation and convection zones, the temperature is 1.5 million degrees C.

After a journey lasting millions of years, the energy finally emerges from the photosphere, the visible surface of the Sun, which has a temperature of about 6,000 degrees C. It is characterized by violent convective motion, intense magnetic

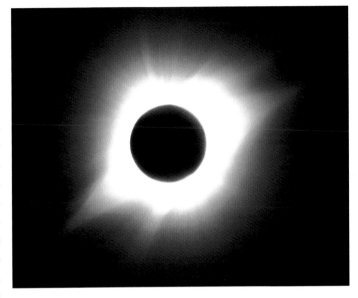

fields and differential rotation rates, whereby areas of the Sun at different latitudes spin at different speeds.

The interaction of these three factors produces structures such as faculae (brighter areas where the magnetic field surfaces, promoting convection and thus heat) and sunspots (areas that appear dark because they are cooler, since the magnetic field is so intense, it blocks the convective motion).

Above the photosphere is the solar atmosphere, divided into the inner part, or chromosphere (less dense than the photosphere), and the outer part, or corona (even more rarefied but stretching out to a distance of five solar radii, as seen in the photo above).

**Above, the solar corona photographed by Fred Espenak of NASA's Goddard Space Flight Center during the total eclipse of July 11, 1991, in Mexico. The corona can be seen with the naked eye only during a total solar eclipse. When sunspot activity is at its maximum, as occurred in 1991, the corona appears to its fullest extent.**

**Right, the solar chromosphere photographed during the total eclipse of November 3, 1994, at Pampa de La Joya, south of Arequipa, Peru. Far right, solar prominences, the largest of which achieved a height of 130,000 km, photographed during the** eclipse of July 11, 1991, at San Blas, Mexico.
■ **These photographs, on Fujichrome 100 and Ektachrome 64 professional film, were taken by Gabriele Vanin with a 100-mm f10 Maksutov lens and exposures of one-quarter of a second.**

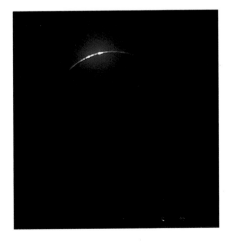

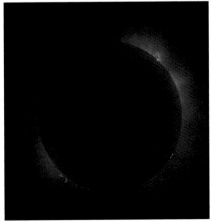

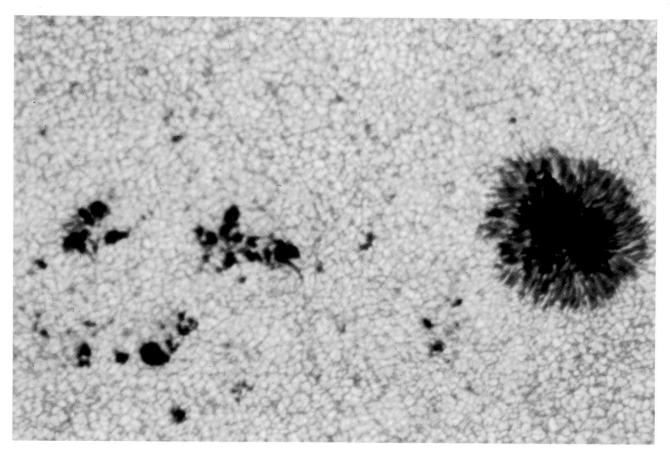

Left, a group of sunspots photographed on June 23, 1989. The central area of a spot, known as the umbra, has a temperature of about 3,600 degrees C. That of the surrounding area, known as the penumbra, is 4,500 degrees C. Visible inside the penumbra are radial structures known as fibrils. Also clearly visible is the granulation caused by the convective motions that affect the photosphere. Each granule is a gas convection cell about 750 km in size that lasts for no longer than 5 or 10 minutes. Sunspots appear where the strongest lines of force in the magnetic field break through to the surface; they tend to form in pairs, often occurring in large groups, like the one shown here, and evolve within a few weeks.

■ A picture taken in visible light with the Mc-Math Telescope, the largest solar telescope in the world (with a 1.5-m mirror and a focal length of 150 m), at the National Solar Observatory on Kitt Peak.

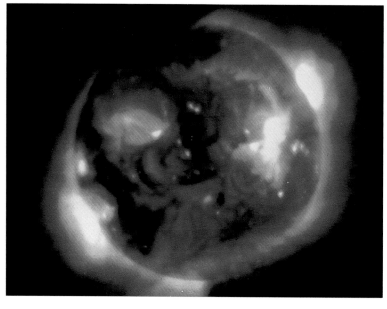

Right, this X-ray photograph of the Sun shows the corona across the entire disk because, of all the outer components of the Sun, the corona is the only one to emit X-rays. Also clearly visible are wide bright patches, which signify areas where the magnetic field is more concentrated, and dark areas, known as coronal holes—semi-permanent features characterized by a total absence of X-ray emissions. Usually present at the poles, such features stretch, during periods of minimum sunspot activity (sunspot activity follows a cycle lasting about 11 years), up to equatorial latitudes and can occupy a considerable percentage of the solar disk.

■ This image was obtained by one of the X-ray telescopes onboard NASA's Skylab space laboratory in 1973.

No fewer than 150 observations of sunspots were made with the naked eye before the invention of the telescope. The largest number came from the Far East, mainly China. Many of the earliest observations, however, are highly suspect, because they could refer to Earthly atmospheric phenomena. This is the case, for example, with an observation by peripatetic philosopher Theophrastus around 350 B.C., as well as a Chinese observation from 165 B.C. The first reliable sighting appears to be one made in 43 B.C., also in China. In the West, records of ancient observations of sunspots are very rare, only about 20, probably a result of the Aristotelian concept of the incorruptibility of the heavens. The first definite sighting dates from 807 A.D., in France. All Western observations up to the time of Galileo were interpreted, according to the prevailing climate of Aristotelianism, not as real features present on the Sun itself but as transits of the inner planets Mercury and Venus across the solar disk. Several observations vie for the privilege of the first telescopic discovery of sunspots: besides Galileo, there were Germany's Christoph Scheiner, Holland's Johann Fabricius and England's Thomas Harriot. It was Harriot who produced the first sketches, on December 8, 1610, which show a mottled Sun.

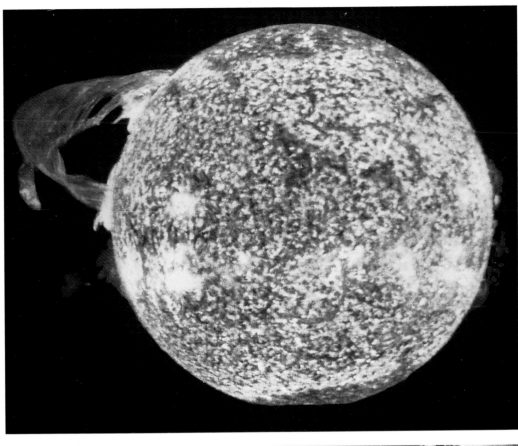

Below, a view of the solar corona in ultraviolet light. The picture has been electronically enhanced in false colors to show the density variations in the corona. Density decreases going from the interior (blue) to the exterior (yellow). Occasionally, flares occur in the solar chromosphere, which are reactions of the solar atmosphere to rapid releases of energy, probably coming from the rapid annihilation of powerful magnetic fields. This release of energy heats the gas and produces an intense electrical field, which, in turn, vigorously accelerates the charged particles. From all this, a temporary heating is produced, localized in the chromosphere (up to 10,000 degrees **C**) and the corona (up to 1 million degrees **C**). These are the most impressive of all the forms of solar activity, during which up to one-tenth of the energy produced by the Sun is released in 1 second. When this occurs, the section of the corona above the flares is completely destroyed and changes shape in just a few minutes.

■ Image produced using data from **NASA**'s Solar Maximum Mission satellite, launched in 1980.

Above, an enormous prominence that rose 420,000 km above the surface of the Sun on December 19, 1973. Prominences are spectacular coronal features, 100 times denser and cooler than the surrounding environment. The Sun's magnetic field perpetuates their duration, which varies from a few hours (active prominences) to several weeks (quiescent prominences). They form as a result of condensation of the corona's matter in a way similar to how clouds form in our own atmosphere. An average quiescent prominence is 200,000 km long, 50,000 km high and 8,000 km wide.

■ Image produced by Skylab's ultraviolet spectroheliograph.

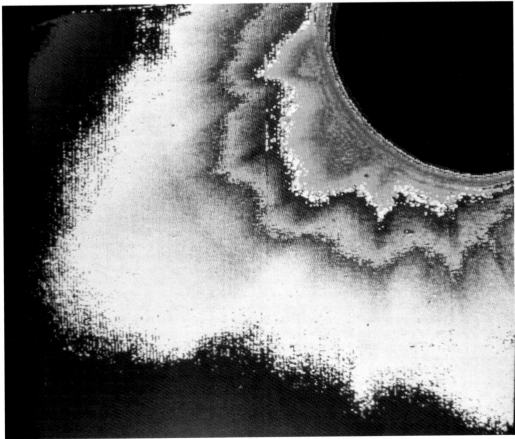

*The first recorded mention of the solar corona may be in Plutarch's* De Facie in orbae lunae *(first century A.D.), where he talks of a "splendor" around the eclipsed Sun. The first definite observation, however, took place in Corfu and coincided with the eclipse of 968. Up to about 1850, it was not fully understood that the corona and the prominences belonged to the Sun and not to a hypothetical lunar atmosphere.*

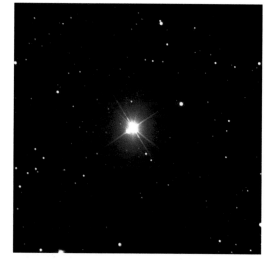

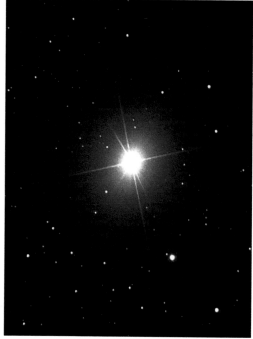

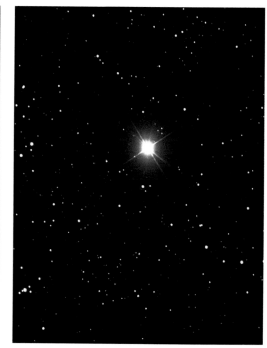

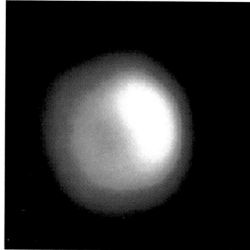

| Proxima Centauri | Barnard's Star | 61 Cygni A | Sun | Alpha Centauri A | Sirius A | Vega |
|---|---|---|---|---|---|---|

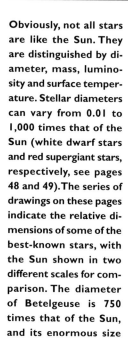

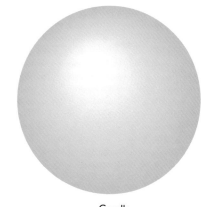

Regulus

Spica

Capella

Obviously, not all stars are like the Sun. They are distinguished by diameter, mass, luminosity and surface temperature. Stellar diameters can vary from 0.01 to 1,000 times that of the Sun (white dwarf stars and red supergiant stars, respectively, see pages 48 and 49). The series of drawings on these pages indicate the relative dimensions of some of the best-known stars, with the Sun shown in two different scales for comparison. The diameter of Betelgeuse is 750 times that of the Sun, and its enormous size makes it the only star

thus far on which it has been possible to glimpse any surface details. These are visible in the false-color image above, taken in red light, at 710 nm, by David Buscher of the University of Cambridge with the 4.2-m William Herschel Telescope in the Canary Islands. The picture is a composite of several brief exposures taken to "freeze" the atmospheric turbulence. The intrinsic luminosity of stars has an even greater range of variation, from a few millionths to millions of times the luminosity of the Sun. The masses of stars, on the

Toward the end of the 17th century, Italy's Francesco Maria Grimaldi carried out experiments in which sunlight was passed through a prism, forming a sequence of colors, from red to violet, known as a spectrum. In 1802, William Hyde Wollaston noticed that the solar spectrum was scored by dark lines. In 1859, Gustav Robert Kirchhoff demonstrated that these lines were due to the absorption of different wavelengths of light by ele-

ments in the atmospheric gases above the surface of the Sun. Because the absorption was different, depending on which gas was responsible, spectral lines at a particular position indicated the presence of that characteristic gas. In time, it came to be understood that the absorption was also defined by the temperature conditions peculiar to each gas. At the same time as Kirchhoff's discoveries, researchers were studying the spectra of oth-

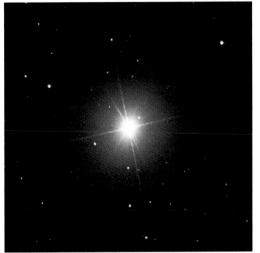

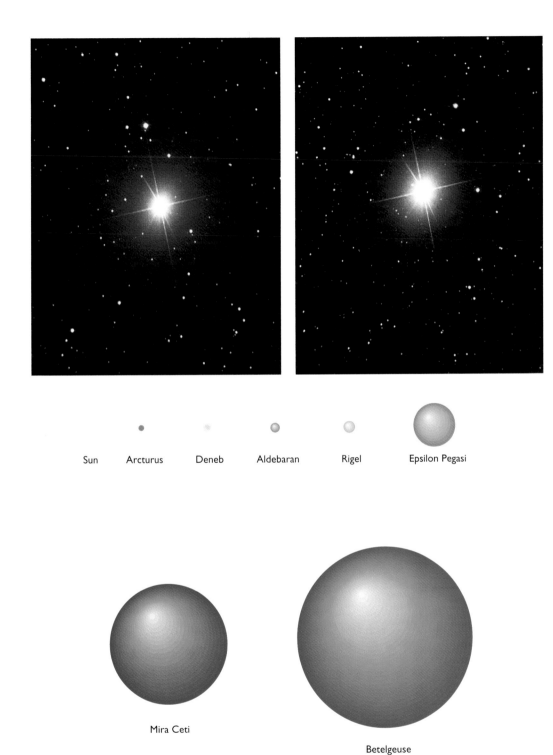

Sun    Arcturus    Deneb    Aldebaran    Rigel    Epsilon Pegasi

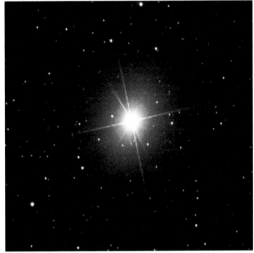

Mira Ceti

Betelgeuse

other hand, vary less, from 0.05 to 100 times the mass of the Sun. Lastly, surface temperatures are from 3,000 to 60,000 degrees C, to which several star colors correspond, from red for the coolest to blue for the hottest, equivalent in turn to different classes, or spectral types, of stars. There are seven main types, known as O, B, A, F, G, K and M, each divided into 10 subtypes, examples of which are shown on these two pages. On facing page (top, from left to right): Mintaka (type O9, temperature 30,000 degrees C), Rigel (B8, 12,000 degrees C), Deneb (A2, 9,700 degrees C).

On this page: Procyon (F5, 6,400 degrees C), Aldebaran (K5, 3,400 degrees C), Betelgeuse (M2, 3,100 degrees C). ■ All photographs are by Gabriele Vanin (6-minute exposures on Scotchchrome 3200), taken with the 200-mm, f5 Newton Telescope at the Observatory of the Associazione Astronomica Feltrina Rheticus (OAAFR).

*er stars. In 1861, France's Jules Janssen identified lines similar to the solar ones in the spectrum of Betelgeuse. Much later, in 1863, Italy's Angelo Secchi was able to classify stars into four main spectral types, with white, yellow, orange and red colors, and to understand that the colors of the stars, their spectral types and their temperatures were strictly correlated. Between 1886 and 1924, a monumental catalog was drawn up at the*

*Harvard College Observatory in the United States, under the direction of Edward Charles Pickering and Annie Jump Cannon, listing 225,000 stars, which were divided, after various attempts at classification, into seven main spectral types plus a few parallel types (variants of type M, known as R, N and S, to which very few stars actually belong).*

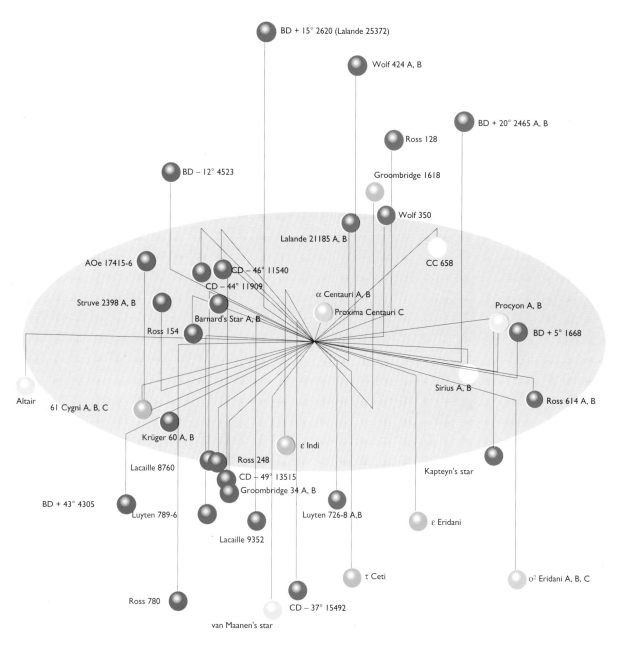

BD + 15° 2620 (Lalande 25372)

Wolf 424 A, B

BD + 20° 2465 A, B

Ross 128

BD – 12° 4523

Groombridge 1618

Wolf 350

Lalande 21185 A, B

AOe 17415-6

CD – 46° 11540

CC 658

CD – 44° 11909

α Centauri A, B

Struve 2398 A, B

Barnard's Star A, B

Proxima Centauri C

Procyon A, B

Ross 154

BD + 5° 1668

Altair

Sirius A, B

61 Cygni A, B, C

Ross 614 A, B

Krüger 60 A, B

ε Indi

Lacaille 8760

Ross 248

CD – 49° 13515

Kapteyn's star

Groombridge 34 A, B

BD + 43° 4305

Luyten 726-8 A,B

Luyten 789-6

ε Eridani

Lacaille 9352

τ Ceti

o² Eridani A, B, C

Ross 780

CD – 37° 15492

van Maanen's star

Left, a diagram showing the positions of the 40 stars that are closest to our Sun. The colors correspond to their spectral types. The presence of letters (**A, B, C**) indicates that the star is double or triple. Distances have been measured trigonometrically, using the diameter of the Earth's orbit as a base. This is how astronomers measure stellar parallax, the small shift in the apparent position of a star against the background stars that results when a star is viewed from opposite ends of the Earth's orbit. The first stellar distance to be measured in this way was that of 61 Cygni. In 1838, Friedrich Bessel found a value of 650,000 times the Earth-Sun distance (the average distance from the Earth to the Sun is known as an astronomical unit, **AU**, equal to 149.6 million km), or 10.4 light-years, which was very close to the actual distance of 11.2 light-years (one light-year is equal to 9.46 trillion km). Of the nearer stars, those that are most like our Sun are Alpha Centauri A, Epsilon Eridani and Tau Ceti, at 4.4, 10.8 and 11.8 light-years, respectively; they are the nearest places outside of our own solar system where one might find other life forms.

Right, Sirius is the brightest star in the night sky and the nearest star visible to the naked eye from the northern hemisphere. It is 8.8 light-years away, putting it actually in seventh place overall, in terms of the nearness of other stars to our Sun. ■ Photographed by Gabriele Vanin with the 200-mm f5 Newton Telescope of the OAAFR (a 5-minute exposure on Scotchchrome 3200). Alpha Centauri, the closest star to the Sun at 4.4 light-years, is actually a triple-star system. In this X-ray image taken by the Einstein satellite, we can see only the two main components: A (very similar to the Sun) and B (slightly smaller and cooler). C is known as Proxima Centauri because it is currently a little closer to us at 4.3 light-years.

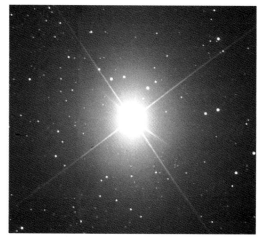

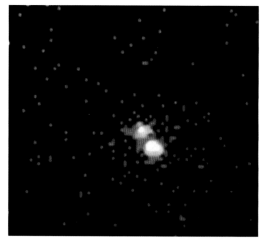

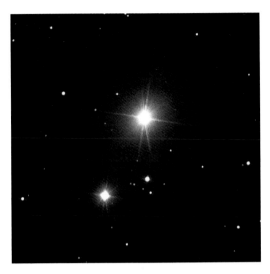

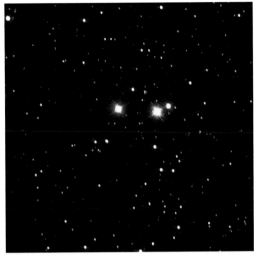

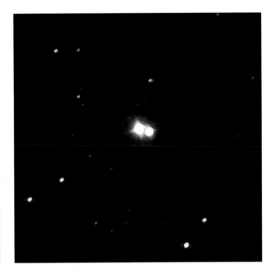

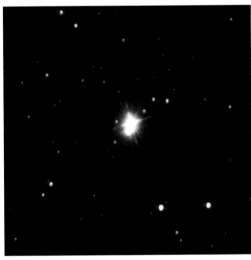

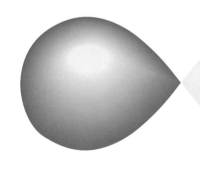

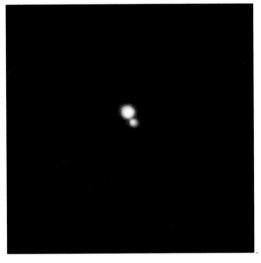

Stars generally occur in clusters and groups, rather than alone. As time goes by, they tend to separate, scattering the clusters. Those that formed at a very short distance from one another, however, may remain gravitationally linked and make up binary, triple or even quadruple (and so on) systems. Half the stars in our galaxy are probably part of multiple systems. In the photographs on this page, we can see four double systems and one triple system (Omicron' Cygni). They are shown in decreasing order of apparent separation, with the first two images magnified 20 times, the following two magnified 40 times and the last one magnified 200 times. Top row, left, Mizar and Alcor, in Ursa Major (10-minute expo-

sure at f5); center, Omicron' Cygni (6 minutes at f5); right, Zeta Lyrae (20 minutes at f10, doubled). Lower row, left, Albireo, in the constellation Cygnus (15 minutes at f10, doubled); right, Gamma Andromedae (10 seconds, with a 9-mm eyepiece-projection adapter). In reality, these systems are sextuple, quadruple, quintuple, triple and quadruple, respectively, but the other companion stars are too faint or too close to the primaries to be visible. The actual separations between the two main stars of each system have respective values of 15,000, 24,000, 2,700, 4,200 and 1,000 astronomical units.

■ These shots were taken with the 200-mm Newton Telescope of the OAAFR on Scotchchrome 3200 film.

Left, some double stars are so close that they can be called "contact binaries": their mutual gravitational attraction distorts them and gives rise to an exchange of matter between the two components.

Double stars have been known since antiquity: Mizar and Alcor are a pair that can be split fairly easily with the naked eye. That Mizar is itself a double star was discovered telescopically: Italian astronomer Giovanbattista Riccioli discovered its companion, Mizar B, which is 50 times closer to Mizar than is Alcor, in 1650. Not all double stars are physically linked. Their apparent proximity may be only an effect of perspective: stars that are widely separated in space may sometimes appear to be

close together if they are lined up from our point of view. It is often impossible to know with absolute certainty what the situation is. There is no absolute certainty of a physical link between the main components in any of the five systems shown on this page, for example (though we can be surer of actual physical links between the fainter companion stars close to the main stars). The first to suggest that binary stars were actually gravitationally connected bodies was English astronomer John

Michell, in 1767, observing the bright star Castor, in Gemini, whose components are fairly close together at about 110 AU (one and a half times the diameter of the orbit of Pluto). Proof of this, however, did not come until 1803, when William Herschel published the results of double-star observations carried out since 1782. These observations showed quite clearly that one of the two components of the Castor system was following an elliptical orbit around the other.

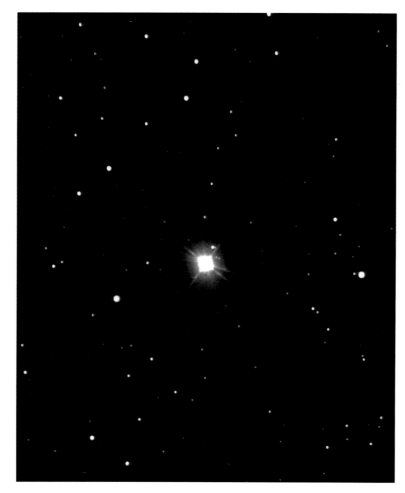

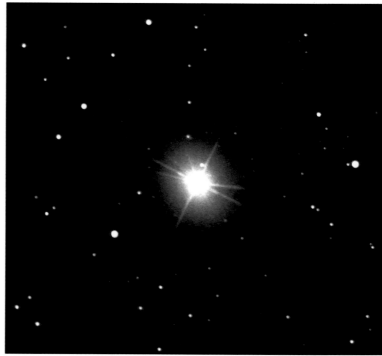

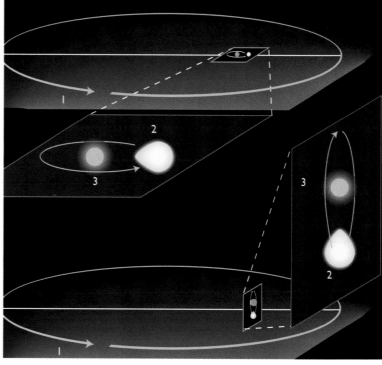

Variable stars are stars whose brightness fluctuates. There are about 40 different types of variable stars. The period of variation ranges from a few hours to decades. Sometimes the variability lies in the very structure of the stars, but sometimes it is caused by external factors, as with eclipsing variables, whose prototype is Algol, in the constellation Perseus.

Above, Algol photograph-ed by Gabriele Vanin at minimum (left) and at maximum (right) bright-ness. The variability is produced by the fact that Algol is a double star. The two components, al-most in contact, have an orbital plane that is al-most edge-on to our line of view. Each time the dimmer companion, a star of spectral type K2 (2 in the drawing) passes in front of the brighter type B8 star (3)—every 2 days, 20 hours and 49 minutes—the light com-ing from the pair is de-creased by about two-thirds. There is also a third companion star, of type F1 (1), which is much more distant, whose or-bit is either parallel or perpendicular to that of the main pair (see dia-gram at right).

■ These shots were tak-en with the 200-mm OAAFR Newton Tele-scope (6-minute expo-sures on Scotchchrome 3200 film).

*The first variable star discovered was Mira, in the constellation Cetus. It was observed for the first time by Dutch astronomer David Fabricius, who took it for a nova. No one noticed it again until 1603, when German astronomer Johann Bayer included it in his atlas. Sometime later, it disappeared mysteriously, reap-pearing after less than a year. In 1638, Phocylides Holwarda of Holland officially recognized its variability, but it was only in*

*1662 that its period and range of variation in brightness were determined. Johannes Hevelius, the great Polish astronomer of the time, suggested the name Mira ("marvelous" in Latin) for this star with its highly strange behavior. Algol's variability was discovered by Italy's Geminiano Montanari in 1667. England's John Goodricke, in 1782, determined its period and also pro-posed that the variation in brightness must be attributed to*

*eclipses produced by a dimmer companion. It was not until 1889 that the hypothesis could be confirmed spectroscopically by Hermann Carl Vogel of Germany, who pointed out the char-acteristic alternation of redshift and blueshift in the spectrum of light coming from the primary star—a pattern produced as the primary alternately recedes and comes closer to us, a move-ment caused by the gravitational influence of its companion.*

Right, Delta Cephei, the prototype of the Cepheid variables, stars that vary in brightness with regular pulsations. An imbalance between the force of gravity (which pulls inward) and the pressure of radiation and gas (which pushes outward) creates instability in a Cepheid's stellar structure. Ionized helium atoms in the Cepheid's atmosphere are ionized a second time by radiation coming from the star's interior and become opaque: the radiation can no longer filter through and exerts a considerable pressure on the atmosphere, making it expand and in-creasing the luminosity and the diameter of the star. As it expands, the atmosphere cools and the helium becomes transparent again, enabling the radiation to flow again and the star to return to its original parameters. Delta Cephei has a period of 5 days, 8 hours, 47 minutes. As can be seen, it is also a binary star: its companion orbits at a distance of no less than 12,500 AU.
■ Photographed by Gabriele Vanin with the OAAFR's Newton Telescope (a 6-minute exposure on Scotch-chrome 3200).

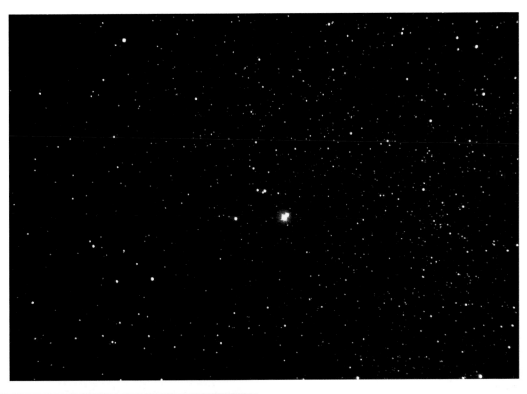

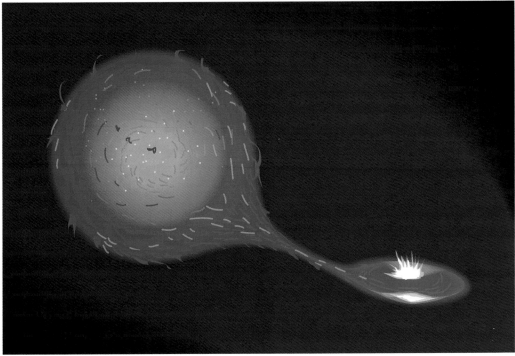

Among the numerous types of variables, many exhibit drastic and sudden changes in brightness. Known as cataclysmic variables, they comprise various subtypes, including novas, dwarf novas and recurring novas. They occur in contact star systems, in which gas is being captured by a white dwarf star from its cooler red companion. The gas forms a growing disk around the white dwarf. In dwarf novas, a subsequent flow of gas from the companion collides with the growing disk at very high speed; the kinetic energy is transformed into heat, creating a highly luminous "hot spot," thereby causing a sudden increase in luminosity. In ordinary novas, the increase in luminosity is much higher and can be attributed to violent nuclear reactions on the surface of the white dwarf inside the accretion disk, following a considerable accumulation of captured gas. Such reactions provoke the expulsion of the outer layers of the star, which, in their turn, generate an expanding shell (see page 130).

*The Cepheids are giant white and yellow stars of spectral types F, G and K, with surface temperatures similar to the Sun's but with a mass 3 to 10 times that of the Sun. Their periods vary from a few hours to 50 days (5 to 8 days on average). These stars assumed a fundamental significance for extragalactic astronomy when American astronomer Henrietta Leavitt, in 1912, discovered a relationship between a Cepheid's period of variation and its absolute magnitude, or intrinsic brightness. Leavitt's observations demonstrated that the longer the period, the greater the brightness. Harlow Shapley, in 1917, calibrated this relationship by using the Cepheids present in globular clusters, whose distance he had been able to measure with mathematical methods. Thus, astronomers obtained a formidable tool for measuring astronomical distances. Once the period of a Cepheid was known, its intrinsic luminosity could be deduced; by measuring telescopically the star's apparent luminosity, it was easy to determine the distance by applying the appropriate formula. Given that the Cepheids are 300 to 40,000 times brighter than the Sun and are therefore visible at great distances, this method can even be applied to measuring distances to other galaxies (see page 135).*

# Nebulas.

Concentrated along the disk of our own galaxy, the Milky Way, are millions and millions of nebulas, large clouds of gas and dust illuminated by the light of hot nearby stars. Emission nebulas, of which the most famous example is M42 in Orion (see photograph at lower right), are mainly hydrogen gas, excited and ionized by ultraviolet light from the stars that are present in the region. When atomic nuclei and electrons recombine, the gas emits light with characteristic wavelengths, among which the strongest is that of hydrogen-alpha, which gives the characteristic red color present in many of the photographs in this book. The human eye, however, is more sensitive to the radiation produced by double-ionized oxygen, which is green in color, and that is why the nebulas appear greenish when seen through a telescope.

Reflection nebulas are made up mainly of dust, which reflects the light of the surrounding stars. Dark nebulas are also made up of dust, but they appear dark because they are very dense and compact and far from any bright stars; they block the light from stars behind them. In a certain sense, nebulas are the wombs in which star clusters are born. In one nebula, there is enough gas to form tens of thousands of stars. Star formation begins with the contraction of the dark globules that can be seen inside many nebulas, known as Bok globules (see photograph on facing page, top right), from the name of the Dutch astronomer who first put forward this hypothesis in the 1950s. Areas of intense star formation, such as the Great Nebula in Orion, glow strongly in infrared because the light of the newborn stars, passing through the dust surrounding the areas of astrogenesis, is absorbed and reddens (see facing page, bottom right).

As diffuse nebulas represent the birthplaces of stars, planetary nebulas and supernova remnants represent their deaths. Planetary nebulas are formed by stars with a mass of 0.8 to 8.0 solar masses. When they have exhausted their hydrogen, the nucleus of such stars contracts and heats, and at the same time, their external layers swell until they become red giants with a diameter tens of times that of the Sun. Following this, an increase in temperature inside the nucleus (up to 100 million degrees C) leads to the fusion of helium and the temporary contraction of the stars. However, when they have finished burning helium, they swell again until they become giants with a diameter hundreds of times that of the Sun. Such a supergiant begins to blow away its outermost layers of gas, creating a fast stellar wind and forming an enormous expanding gas bubble. Meanwhile, the star's surface increases in

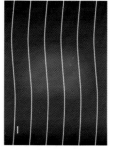
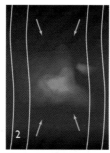
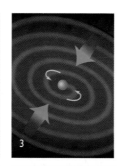
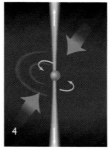
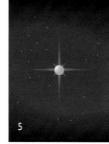

The various stages of gravitational contraction of gas leading to the birth of a star: (1) Initially, the collapse of the gaseous mass is prevented by the presence of magnetic fields. (2) Afterward, however, the fields tend to weaken, leaving the nucleus of the cloud at the mercy of the force of gravity, which flattens it, forming a protostar. (3) From the gaseous material that falls toward the interior, the protostar forms an accretion disk around itself. (4) The process of accretion produces jets that radiate out from the disk. (5) A star is truly born, and from the material remaining in the disk around it, embryonic planets begin to form. (6) Finally, these become true planets, and a new solar system is born.

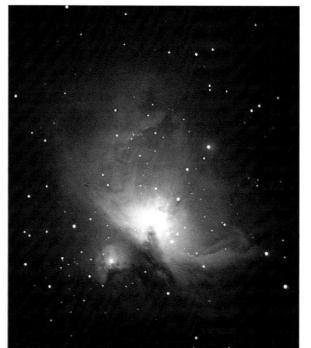

The brightest and best-known emission nebula is M42, the Great Nebula in Orion.
■ Photographed by Gabriele Vanin using a 254-mm f10 Meade Schmidt-Cassegrain telescope and f6.3 Celestron field corrector (a 10-minute exposure on Scotchchrome 3200).

temperature, from 25,000 to 200,000 degrees C, and it gradually ionizes increasingly wider areas of the expanding gas, which begins to shine through fluorescence; and thus, a planetary nebula appears. The gas expands at a speed of 5 to 100 km per second. The planetary shines until, continuing to expand, it becomes too large and diffuses into the interstellar space. This occurs 10,000 to 30,000 years later. The central star, on the other hand, cooling progressively, eventually becomes a white dwarf, a highly compact star containing a mass similar to that of the Sun in a volume only as large as the Earth. Larger stars of 8 solar masses or more, on the other hand, continue to contract, reaching a temperature of 5 billion degrees C, allowing the formation of increasingly heavier elements, such as carbon, oxygen, silicon and, finally, iron. The nucleus of the iron atom is the most stable in nature, and this stops the transmutation of the ele-

ments. At this point, the nucleus of the star, contracting again, reaches a temperature of 8 billion degrees C. A cycle of highly violent impacts between the nuclei of the various elements present is unleashed; the iron nuclei are heavily bombarded and react by transforming themselves not into heavier elements but into helium nuclei. This reaction generates a powerful absorption of energy; the equilibrium of the star breaks down, and it finishes its life with the stupendous explosion of a supernova, during which at least half (or even the entire mass) of the star is scattered into space. In the first case (Type-II supernovas), a black hole remains, or else a neutron star. In the second case (Type-I supernovas), only the supernova's ashes remain. The supernova remnant shines through the same phenomenon of fluorescence as described earlier, fueled by the energy released during the blast. Because of the greater mass involved, the supernova remnant

lingers for a longer time before dissipating. The speed of expansion is tens of thousands of kilometers per second.

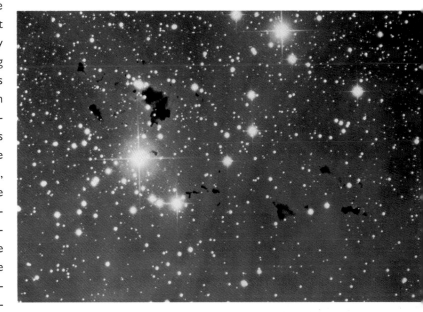

Bok globules, star-forming regions within nebulas, stand out clearly in this picture of **IC 2944**, an emission nebula.
■ Tricolor photograph by David Malin, using three exposures (25, 25 and 35 minutes) taken with the Anglo-Australian Telescope in blue, green and red light by David Malin and Steve Lee.

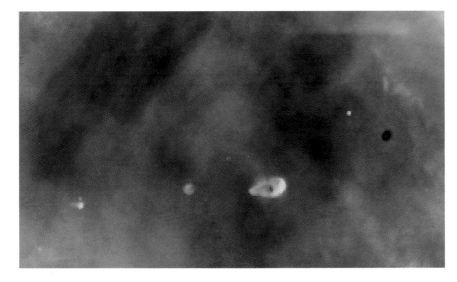

A close-up of part of the area of the Orion Nebula shown on page 135 reveals, thanks to the **Wide Field/Planetary Camera 2 (WF/PC-2)** on the Hubble Space Telescope, five developing stars around which there are in all likelihood protoplanetary disks, that is, newborn solar systems **(C.R. O'Dell/ Rice University-NASA).**

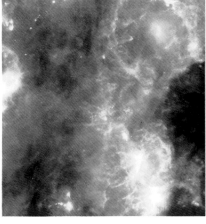

The strong infrared emission of the region between Orion and Monoceros, visible in this picture taken by **NASA's Infrared Astronomical Satellite (IRAS),** indicates a considerable amount of star formation taking place. The two bright spots visible at lower right and on the left are the Orion Nebula and the Rosette Nebula, respectively.

Right, the Lagoon Nebula, in Sagittarius, is so called because of the presence of the dark swath that crosses it on one side. It contains many Bok globules and luminous regions, which in this picture emerge in their minutest detail, thanks to the use of the technique of unsharp masking. The nebula's distance is about 2,500 light-years, and its dimensions are 30 by 65 light-years.

■ Photograph by David Malin, obtained by combining three plates exposed by Sue Tritton and John Barrow in blue, green and red light on the 1.2-m United Kingdom Schmidt Telescope (UKS), using exposures of 40, 60 and 60 minutes.

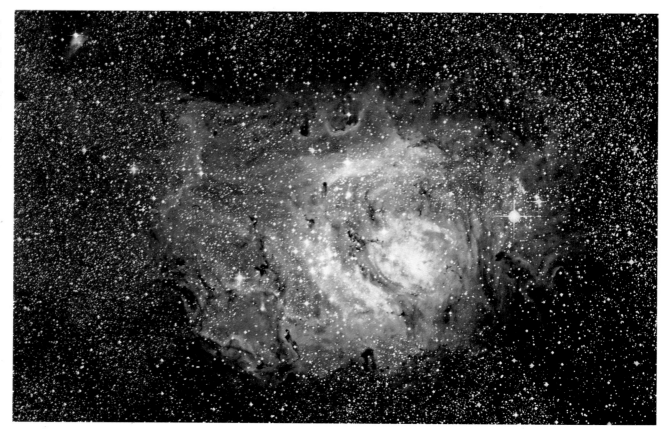

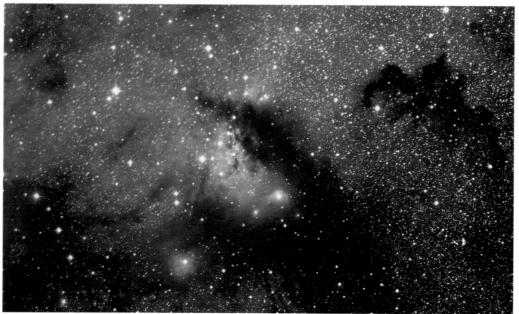

*The North America Nebula was discovered photographically by German astronomer Max Wolf in the early 1880s. He called it the "America Nebula." Its actual, and more appropriate, name was suggested in 1903 by American astronomer Edward Emerson Barnard. This nebula appears to have increased in brightness over the years: up to 1950, it could be observed only photographically, but now, it can be seen with small telescopes or*

*binoculars, or even with the naked eye, under conditions of good sky transparency.*

*The Lagoon Nebula is easily visible with the naked eye on a very clear night, appearing as a fuzzy patch just to the right of the main band of the Milky Way. Although its discovery is generally attributed to France's Le Gentil (in 1747), it was probably seen first by British Astronomer Royal John Flamsteed in 1680.*

Left, NGC 6188, in the constellation Ara. In this region, red emission nebulas, blue reflection nebulas and dark nebulas (produced by dust that is not illuminated by any star) are visible.

■ David Malin created this image by combining shots taken with the 3.9-m Anglo-Australian Telescope (AAT), using (as is the case for most of his pictures reproduced here) IIa-O plates with a GG 385 filter (blue light), IIa-D with a GG 495 filter (green light), 098-04 with an RG 610 filter (red light) and exposures of 30, 25 and 35 minutes.

Facing page, the North America Nebula lies 3 degrees east of the bright star Deneb, in the constellation Cygnus. For a long time, it was thought to be illuminated by Deneb and therefore at the same distance as Deneb (1,600 light-years). Recently, however, it has been discovered that the source of illumination is a very bright star that is made almost invisible by the absorption of its light by interstellar dust, which weakens it by 100 million times. The distance of the nebula is 450 light-years, and its dimensions are 13 by 15 light-years.

■ Tony and Daphne Hallas of Oakview, California, produced this composite image from two Fujicolor HG 400 negatives exposed for 70 minutes each with a 15-cm f7.5 EDF Astro-Physics refractor.

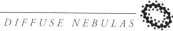

Below, a dark cloud in Scorpius, seen in its actual colors for the first time in this picture by David Malin. Here, the dark, dusty matter is contained in a nebula swept by very strong stellar winds coming from a nearby cluster of very hot stars, NGC 6231 (out of the field), located in the heart of a group of young stars known as the Sco OB1 group. The ultraviolet radiation of these stars is responsible for the halo of red light that surrounds the dark cloud. Inside the globule, a few stars are immersed, whose light feeds a few reflection nebulas, blue-white in color. The cloud is 6,000 light-years away, and its dimensions are 25 by 15 light-years.

■ David Malin produced this image from three plates exposed in blue light (for 19 minutes), green light (25 minutes) and red light (35 minutes) at the prime focus of the AAT. The blue plate was taken by Malin himself, the other two by Steve Lee. The plates were then combined, using the process of photographic amplification.

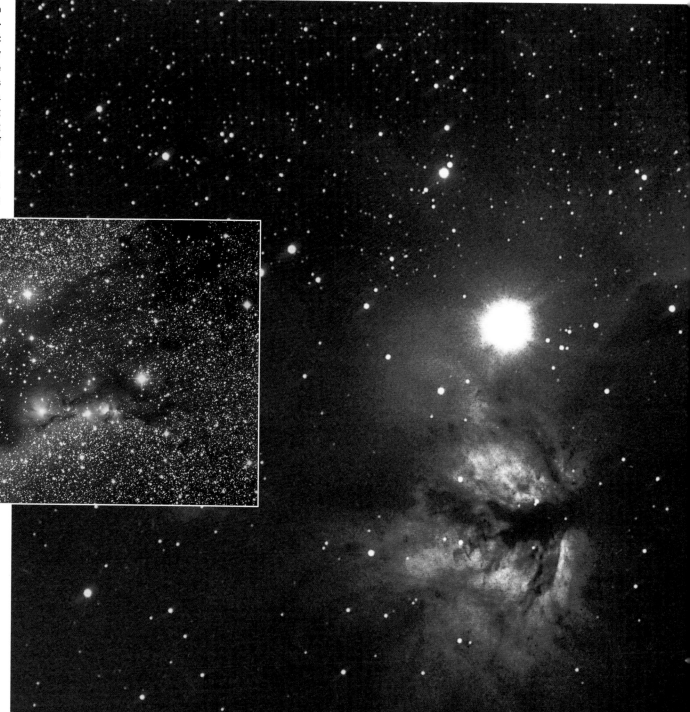

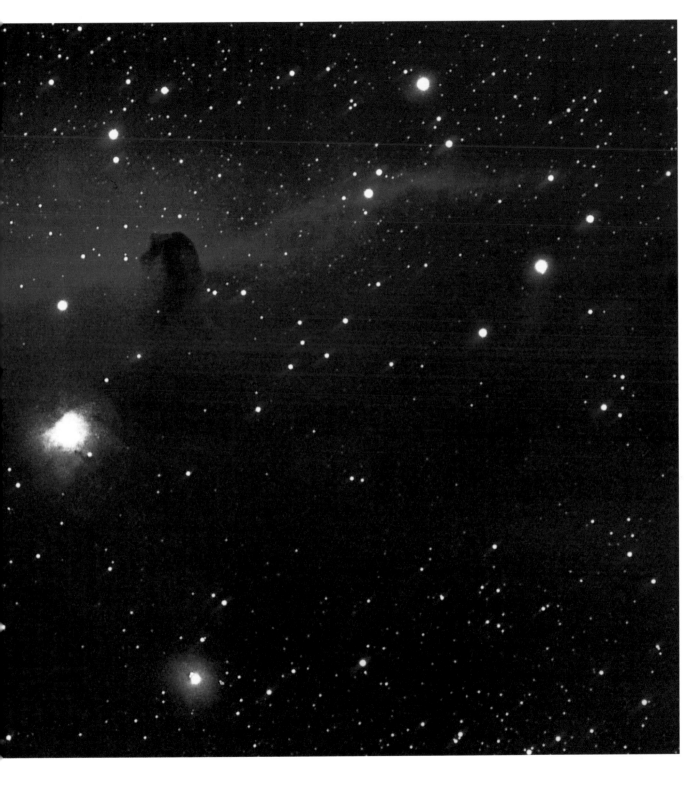

The Horsehead Nebula is the most prominent feature of a large dark cloud, formed of dust particles covered with water molecules and carbon dioxide, in **IC 434.** The gas which comprises the latter is illuminated by the bright star Sigma Orionis (out of the field), while the dust remains cold and inert to the radiation. The distance of the nebular complex is 1,650 light-years, and its dimensions are 30 by 10 light-years. Alnitak, the bright star seen in the left half of the picture, is actually much nearer to us, at 1,100 light-years. Under it appears the nebula **NGC 2024,** illuminated by the radiation of invisible stars hidden within the cloud itself, which also contains condensations that could be stars in their earliest stages of life.

■ Tony and Daphne Hallas created this picture using two hypersensitized 120-format Fujicolor 400 **HG** negatives, each exposed for 85 minutes with a 15-cm f7.5 Astro-Physics **EDF** refractor. The exposures were guided automatically using an **SBIG ST-4 CCD** camera. The negatives were then combined in the darkroom.

*The Horsehead Nebula is the most famous example of a dark nebula and is one of the best-known objects in astronomy. The red fog surrounding it, known as IC 434, was discovered photographically by New Zealand-born American astronomer William Pickering in 1889, while the dark cloud was seen on a plate taken in 1900 by English astronomer Isaac Roberts, who described it as a "bay." It was E.E. Barnard who first interpreted it as a large dark mass projected against a bright gaseous background. For a long time, no one was able to detect it visually. Barnard failed to do so in 1913, even with the Yerkes 1-m refractor. However, after several attempts, all in the USA, Walter Scott Houston and Leslie Peltier succeeded in glimpsing it with instruments that had apertures of only 12 to 15 cm. Even now, seeing it without using a filter is a real challenge.*

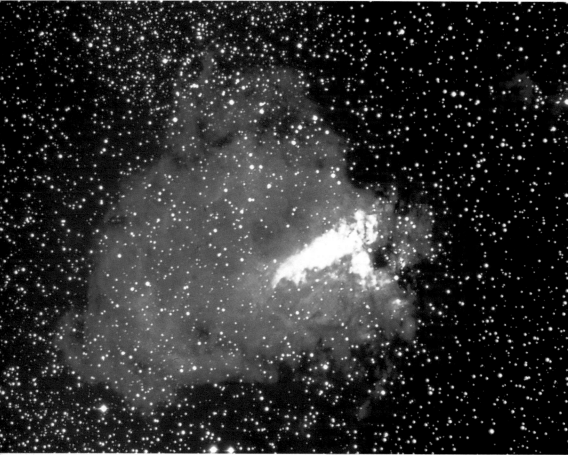

Facing page, the Trifid Nebula, in Sagittarius, so named because it is divided into three sections. It lies in the same region as the Lagoon, at the same distance. It is about 20 light-years across. At its center is a compact group of stars, of which one very bright star appears to produce most of the light illuminating the Trifid's red (emission) and blue (reflection) gas clouds.
■ Tricolor composite by David Malin, using plates taken by Dave Jauncey with the AAT (exposures of 35, 30 and 30 minutes).

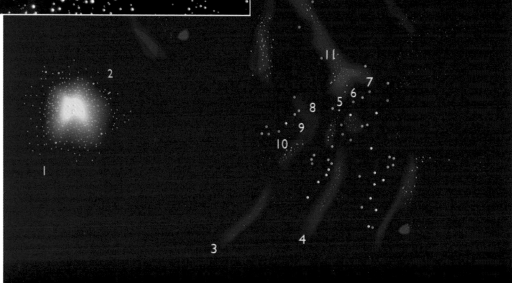

Above, the Omega Nebula, in Sagittarius, is 5,700 light-years away, and its dimensions are 60 by 75 light-years. Its total mass is enough to form 800 stars as large as the Sun. The nebula's main source of illumination lies in a small group of stars nestled at its center. In the hottest sector of the nebula, the intensity of the gas's spectral emission lines is such that the resulting visual impression is of a vivid white color.
■ Tricolor composite image produced from three black-and-white plates exposed in blue, green and red light by Richard West using the ESO's 3.6-m telescope at La Silla, Chile.

Right, an illustration showing the distribution within the Milky Way of most of the nebulas discussed in these pages. Our solar system is included, located at a distance of about 30,000 light-years from the galactic center, between the two spiral arms known as Sagittarius and Orion.

1) Galactic nucleus
2) Nuclear bulge
3) Sagittarius arm
4) Orion arm
5) Sun
6) Orion Nebula
7) Rosette Nebula
8) Lagoon Nebula
9) Trifid Nebula
10) Eagle Nebula
11) Eta Carinae Nebula

*The Omega Nebula was discovered by Swiss astronomer de Cheseaux in 1764. He wrote: "... it has an altogether different appearance from the others; it has the perfect shape of a ray or the tail of a comet ... its sides are extremely parallel... the center is whiter than the edges." The great William Herschel, discoverer of Uranus and founder of stellar astronomy, described it in 1785 as a wonderful patch of nebulosity with a homogeneous appearance.*

*The Trifid Nebula was probably first observed by Le Gentil in 1747. William Henry Smyth, the famous English amateur astronomer, saw it toward the middle of the 19th century as "a particular glare surrounding the delicate triple star placed at the center, where it opens out...." It is believed to have been John Herschel, son of William, who was the first to call it by its present name, also around the mid-19th century.*

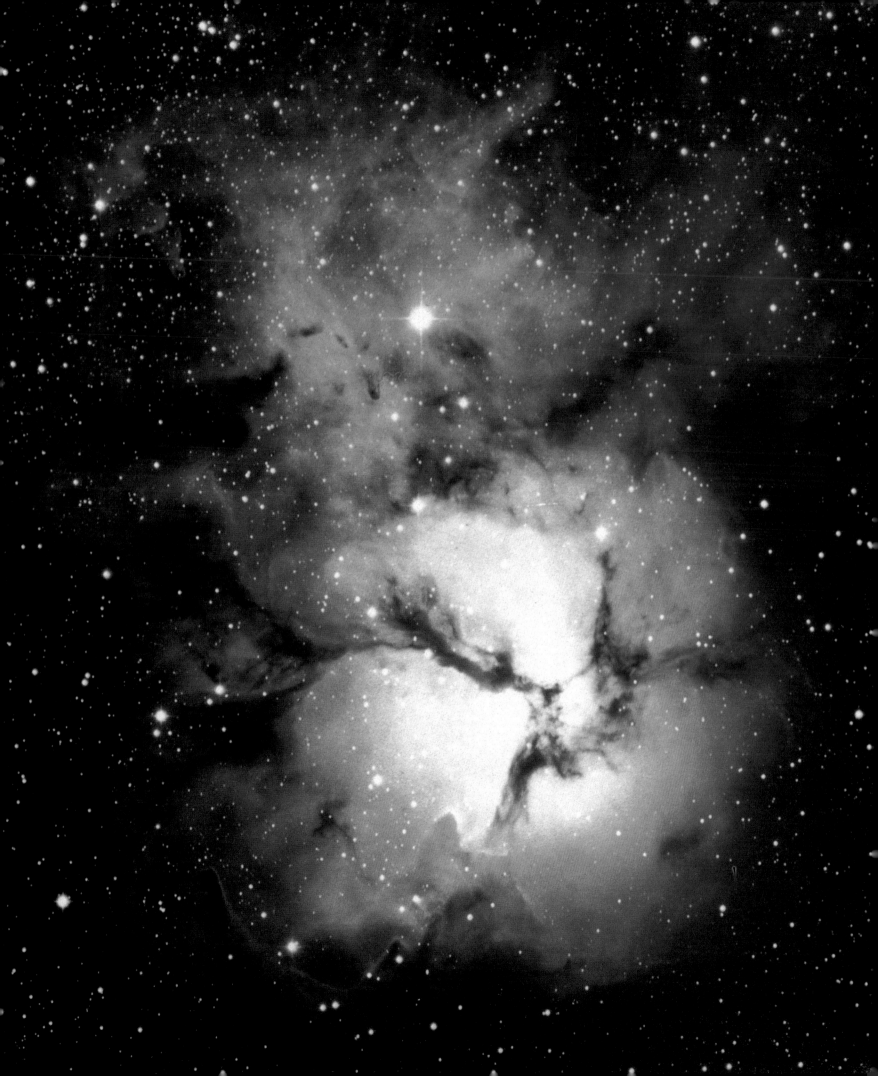

NGC 6334, known as the Red Nebula, lies in the constellation Scorpius. It is rather small but very complex. It is situated at a distance of 5,700 light-years, and its dimensions are 65 by 50 light-years. Its red coloration, which is mainly due to emission at the wavelength of hydrogen-alpha, is enhanced by differential absorption by the interstellar dust lying between us and the nebula. The nebula is situated on the galactic equator, where the concentration of interstellar dust is most intense. The dust acts like a kind of filter that absorbs most of the blue light and lets most of the red through. Although we are thus unable to see blue stars in the nebula, careful observations of radio waves have shown that there are in fact many hot, bright stars, which are responsible for most of the emission of the cloud.

■ Tricolor composite image from three black-and-white plates exposed in blue, green and red light (for 45, 45 and 60 minutes) by Richard West using the ESO's 3.6-m telescope. As with most of the tricolor ESO images appearing in this book, this picture was produced using IIa-O, IIa-D and 098-04 plates and GG385, GG485 and RG630 filters.

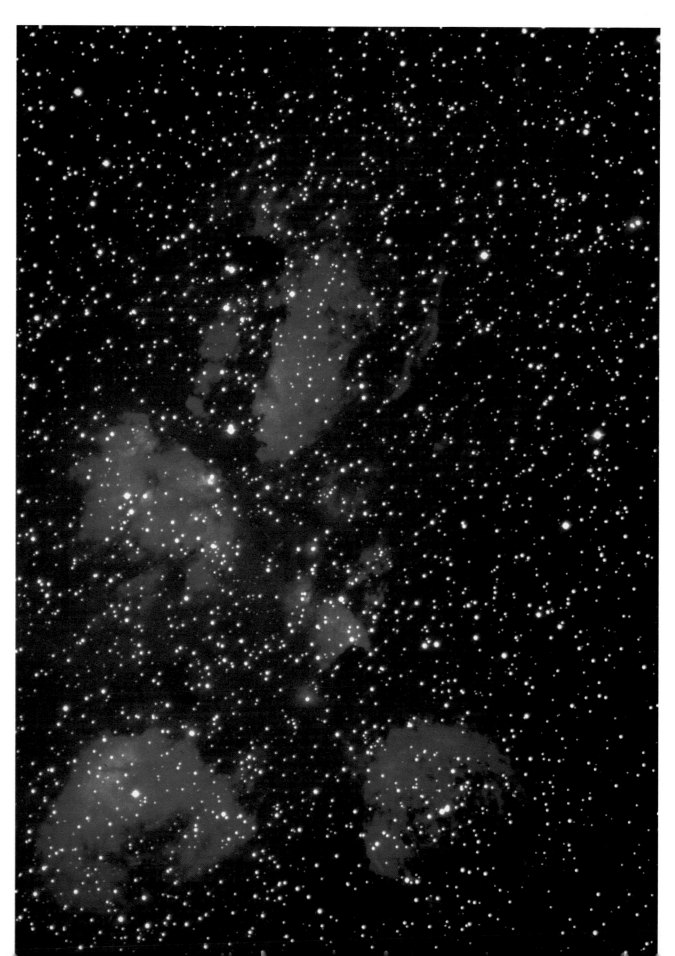

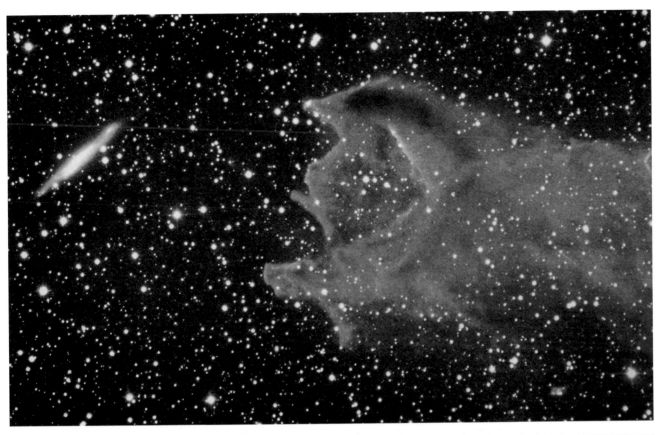

The cometary globule CG 4 is an example of an isolated dust cloud that is about to be destroyed by radiation coming from the star Zeta Puppis (out of the field), which sweeps the nebula from left to right. The "tail" of the globule appears blue because this is a reflection nebula. However, near the globule's head, where light is seen through the dust, the color turns green and, here and there, yellow and brown. The red is due to the hydrogen left in the nebula following the luminous bombardment.

■ Photograph by David Malin, using plates exposed by Despina Hatzidimitroiu at the AAT for 35, 30 and 35 minutes.

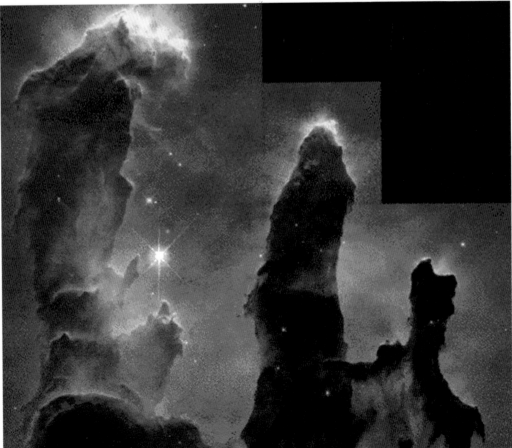

The Eagle Nebula in Serpens is 7,000 light-years away and 70 by 55 light-years wide. This extraordinary image of its central region has an incredible three dimensionality. The nebula's luminosity is produced by ultraviolet radiation from a cluster of hot young stars known as M16, which is part of a larger group of evolving stars. The radiation causes gas in the dark turrets visible in the photo to evaporate, exposing the gaseous globules, or nascent stars, that are emerging from the parent cluster.

■ The image is composed of three shots taken in red, green and blue light with the Hubble Space Telescope's WF/PC-2.

Right, the central part of the nebula NGC 3372, which surrounds the star Eta Carinae, lies 8,000 to 9,000 light-years away. At the heart of the nebula is a group of very hot, young stars, among the brightest in the galaxy, about 1 million years old. There are also several other young star clusters. The entire cloud has a diameter of 300 light-years.

■ Tricolor composite image produced from plates exposed by S. Laustsen and J. Surdej in blue, green and red light (for 5, 15 and 15 minutes) using the ESO's 3.6-m telescope.

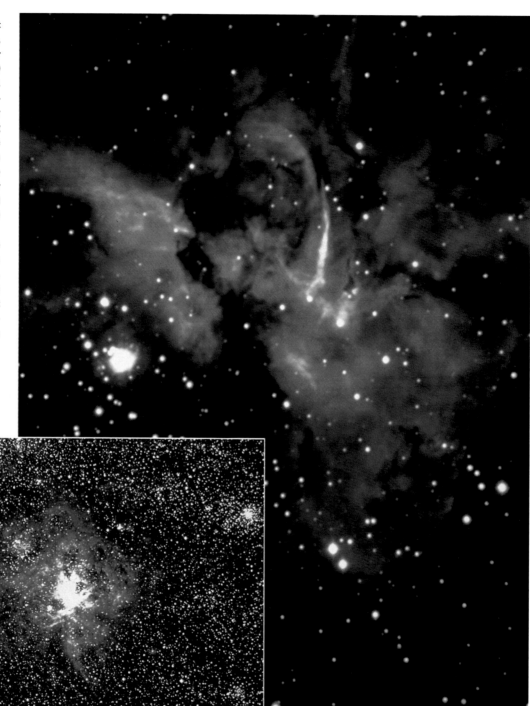

Below left, the Tarantula Nebula, located in the Large Magellanic Cloud, a satellite galaxy of the Milky Way. The nebula's unusual name is derived from its appearance when observed visually through a telescope. It was depicted in a drawing executed in 1870 by M. Le Sueur with the 122-cm Grubb reflecting telescope at Melbourne. The center of intense star-forming activity, its dimensions are colossal, about 650 light-years across. It contains enough matter to form half a million stars like the Sun. At the center of the Tarantula Nebula, there is a compact object whose highly powerful radiation ionizes and illuminates the gas of the surrounding nebula. The object, 30 Doradus, was for a long time considered to be a single star. High-resolution pictures taken from Earth and from space have, however, shown that this is a cluster of thousands of very hot, young stars (see page 134).

■ Tricolor composite of three plates exposed in blue, green and red light (for 60, 45 and 60 minutes), taken with the ESO's 1-m Schmidt telescope by G. Pizarro.

Eta Carinae, the bright star in the center left portion of the photograph above (top), has a very unusual history. In 1677, the great Edmond Halley observed it as a 4th-magnitude star. In 1843, it became the second-brightest star in the sky, after Sirius. (In fact, in terms of intrinsic luminosity, Eta Carinae was at that time actually the brightest star in our galaxy—600,000 times more luminous than Sirius and 12 million times brighter than the Sun. It appeared less bright than Sirius only because it is so much farther away.) Over the following years, it decreased in brightness, becoming invisible to the naked eye. Today, we know that its former brightening was the result of a violent expulsion of a considerable portion of the star's atmosphere, which produced a veritable explosion of light. Subsequently, the outer portions of the star's atmosphere cooled, becoming more opaque, thus diminishing its brightness. It now appears to be almost completely shrouded by a cloud of gas (see also page 134).

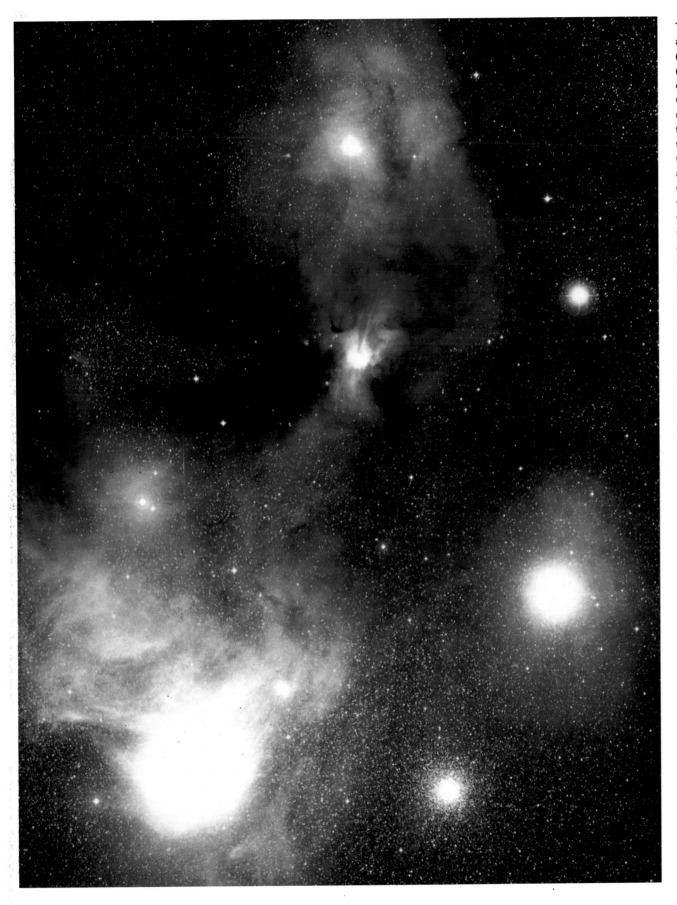

The reflection nebulas around **Rho Ophiuchi** (at top) and **Antares** (lower left). Clouds of dust particles reflect the light of the surrounding stars. The difference in color of the two clouds surrounding the main stars is due to a real difference in the stars' surface temperatures and, therefore, their colors. Rho Ophiuchi is a quadruple star, all of whose components are blue and very hot, with surface temperatures of 15,000 to 25,000 degrees C, while Antares is a cool red supergiant (which, however, appears yellow-orange) with a temperature of 3,000 degrees C. The dust molecules around Antares came from the star itself, from reactions involving elements ejected from the core by the powerful stellar wind.

■ Tricolor composite produced by David Malin, using plates exposed with the **UKS** by Sue Tritton and John Barrow for 60 minutes.

The region comprising the emission nebula **IC 1283-84** (upper right) and the reflection nebulas **NGC 6589** and **NGC 6590** (lower left). The stars hidden by the two bluish patches of **NGC 6589** and **NGC 6590** do not have enough energy to generate the ultraviolet light necessary to constitute an emission nebula, but they are sufficiently luminous for their light to be reflected by the dust cloud toward which they are drifting. The star nestled at the heart of the **IC 1283-84** nebula, on the other hand, is bright enough to excite the surrounding region, rich in hydrogen as well as dust. The mingling of the blue light reflected by the dust with the red light produced through the fluorescence of hydrogen gives the whole a delicate magenta color. ■ David Malin produced this image using three plates exposed in blue, green and red light with the **AAT** for 35, 30 and 40 minutes, respectively.

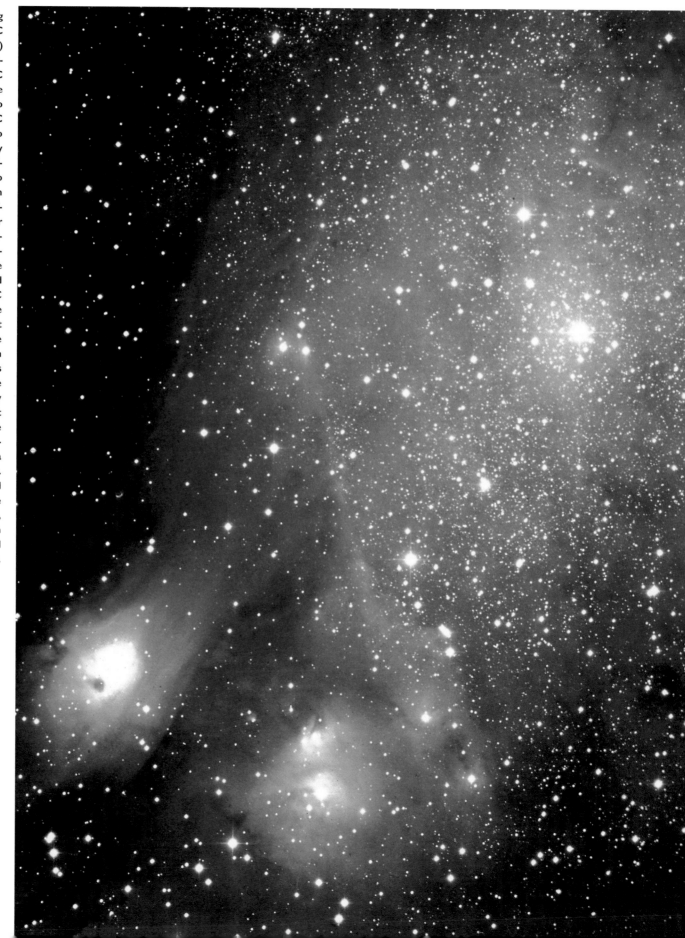

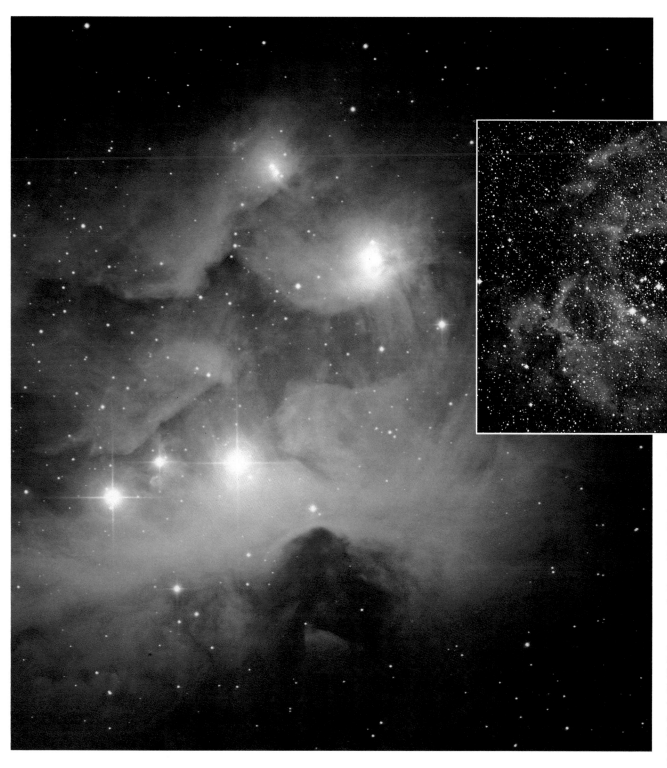

Left, the marvelous reflection nebula surrounding **NGC 1973, NGC 1975** and **NGC 1977**, star clusters just north of the Orion Nebula. Inside the cloud, there are cavities that may have been produced by stellar winds, in which the fluorescence of residual hydrogen produces a reddish glow.
■ **Photograph by David Malin, a tricolor composite of three plates exposed at the AAT for 15 minutes each.**

Above, the Rosette Nebula in Monoceros. Out of its gases, the star cluster **NGC 2244** formed. The winds of the young stars swept away the gas, creating the central cavity, 12 light-years in diameter. It is expanding at 20 km/sec, and in a few million years, the gas will disappear.
■ **Photographed by William Miller (a 3-hour exposure on Super Anscochrome 100) with the 122-cm Schmidt telescope on Mount Palomar.**

*The first photograph of a nebula was taken by the American doctor and amateur astronomer Henry Draper, on September 30, 1880. He took a picture of the Orion Nebula with a 28-cm Clark refractor from his observatory at Hastings on Hudson, 30 km north of New York. The exposure time was 51 minutes on a newly conceived plate of silver bromide gelatin, much more sensitive than others in use at the time. In spite of this, the gas of the nebula was barely visible. Draper improved on this, the same year, with an exposure of 104 minutes and, in 1882, with another of 137 minutes, in which the gas was very clearly visible. In 1883, English astronomer A.A. Common was able to obtain a picture of M42 very similar to those of today, with just a 37-minute exposure and a 1-m reflector.*

Almost all stars of modest mass, having exhausted the helium at their core, swell and begin to blow very strong winds that carry away some of their mass and form extensive shells, the planetary nebulas.

At right (top) is a newly formed planetary nebula, Hen 1357, photographed by Matt Bobrowsky (CTA) with the Wide Field/Planetary Camera 1 (WF/PC-1) of the Hubble Space Telescope. Analysis of the details visible will help explain many stages in the formation of these systems.

At right (below), the newly formed planetary nebula Hen 1357, 18,000 light-years distant, photographed by the Space Telescope's WF/PC-2. It is barely one-tenth the diameter of a normal planetary nebula and is thought to have formed only 20 years ago. The tiny central star is reduced to a core of carbon and oxygen at a temperature of 70,000 degrees C, which is becoming a white dwarf.
■ Photograph taken with the KPNO's 3.8-m reflector.

The diagrams below show the reactions that occur in stars similar in size to the Sun at a temperature of 100 million degrees C, after all the hydrogen has been exhausted. Three helium nuclei combine to form one carbon nucleus, which, combining with a helium nucleus, forms one oxygen nucleus (top).

The reactions that occur in more massive stars, at temperatures of 600 million to 1 billion degrees C, generate more complex nuclei. From the fusion of two carbon nuclei, first magnesium and then sodium and neon are generated. From the fusion of two oxygen nuclei, sulfur is produced, followed by phosphorus and silicon (bottom).

Facing page, images of four planetary nebulas taken with the Hubble's Wide Field/Planetary Camera 2. At right (top), the Egg Nebula, taken in red light, shows the evolution of a giant red star into a planetary nebula. The brilliant arcs are dense puffs of material emitted by the central star and expanding at a velocity of 20 km per second. These luminous jets indicate areas where light escapes more easily because the haze that encircles the dying star is more diffuse there.

Far right (top), a composite image of NGC 7027 from two images taken in visible and infrared light. This image shows that in the early phases of nebula formation, the expulsion of outer stellar layers is slow and discontinuous and follows a spherical profile. Later, a vigorous nonspherical eruption occurs (illustrated by the brilliant central regions in the image), and dense clouds of dust condense from it.

At bottom, two planetary nebulas with a very complex and characteristic appearance (composites from images in red, green and blue light). At lower left, NGC 6543, known as the Cat's Eye Nebula; at far right, MyCn18, nicknamed the Hourglass Nebula. The distinctive twin-lobed appearance of these two nebulas could be the result of a swift stellar wind emitted by the central star and expanding more slowly and thickly at its equator than at its poles.

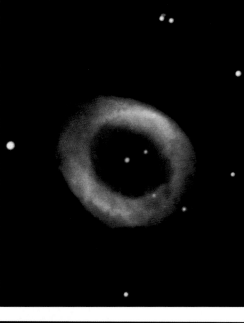

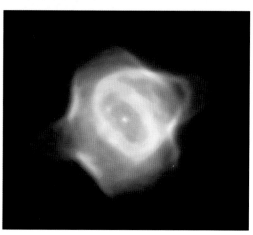

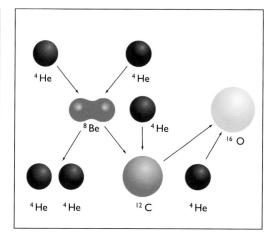

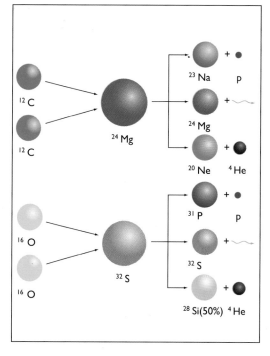

*The Ring Nebula in Lyra was the first planetary nebula to be discovered. Its discovery is attributed to the French astronomer Antoine Darquier in 1779. Observing through an 8-m telescope, he described it as "a perfectly outlined disk as large as Jupiter, but with an opaque light, like a planet seen through fog." William Herschel, in 1785, included it among "curious objects in the sky." The nebula's central stars are 15th-magnitude, 4,000 times fainter than the faintest stars visible to the naked eye. They were observed for the first time by German astronomer F. von Hahn, around 1800.*

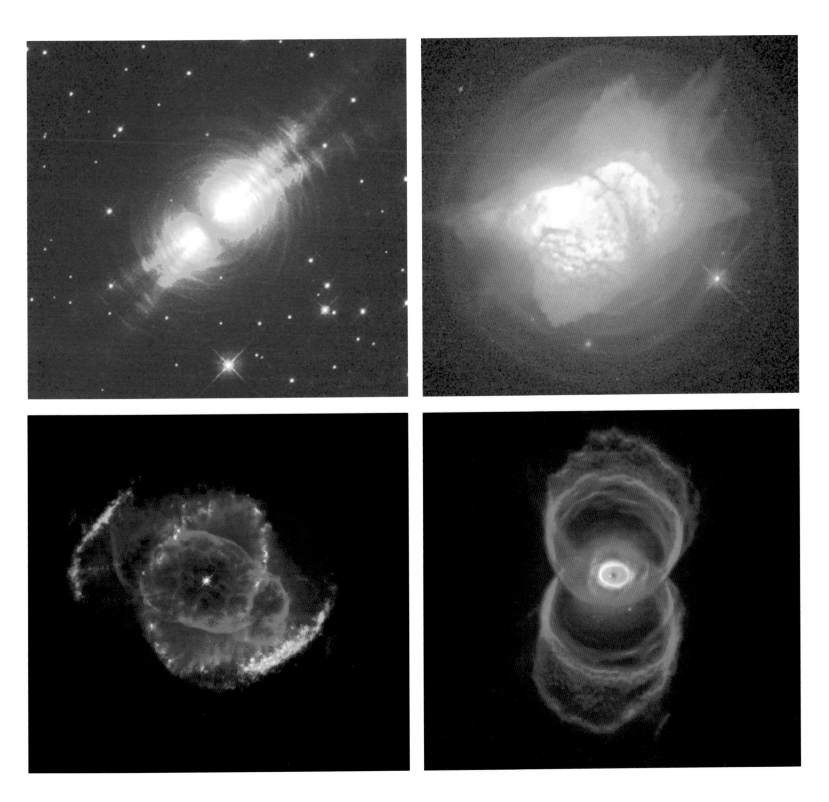

William Herschel, in an article written in 1785, coined the term "planetary nebula" for these bodies, whose spheroid shapes made them resemble planets. Here is his description of one of them: "A curious nebula, I do not know what else to call it. It has a slightly oval, almost circular shape, and magnified 460 times, it appears about 10 or 15 seconds in diameter... its brightness does not change much with magnification, as if it were a body which is planetary in nature; however, it seems to be of a stellar type." Herschel always showed a planetary nebula through his telescope to any visiting foreign astronomer, asking them to pronounce on the nature of the object. According to Herschel, planetary nebulas constituted one of the initial phases of star formation, an intermediate stage between the creation of diffuse nebulosity and subsequent condensation into stars.

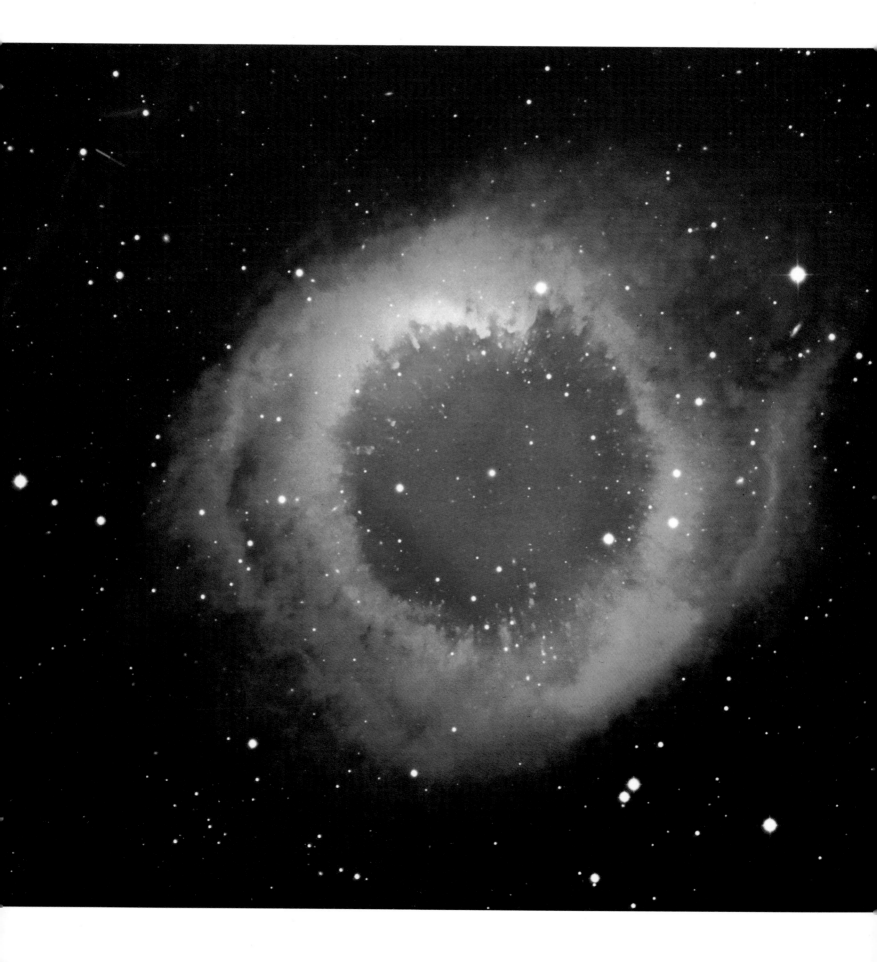

Left, the Helix Nebula, **NGC 7293**, in Aquarius, is the nearest planetary nebula, at 300 light-years. Visible at the center is the generating star, which has now become a white dwarf. Planetary nebulas often have a ringlike appearance because our line of sight passes obliquely through the edge of the shell, thus meeting more matter there than at the center.

■ Tricolor composite by D. Malin, using plates exposed by him at the AAT for 35, 35 and 90 minutes.

Right, the planetary nebula **NGC 3132**, in Vela, is 2,600 light-years away. The color of planetary nebulas tends to be blue-green in the central part (where most of the energy coming from the star manages to ionize oxygen) and red at the edges (where, at lower temperatures, only hydrogen is ionized).

■ Tricolor composite of three plates taken with the ESO's 3.6-m telescope.

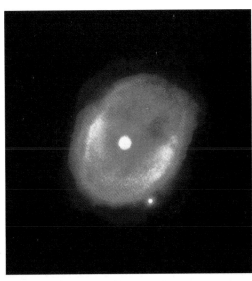

Below, the Dumbbell Nebula, **M27**, in the constellation Vulpecula. Its shape is irregular because the expanding gas has encountered greatly differing degrees of resistance to its penetration of the interstellar medium. The cloud's gases are expanding at the rate of 27 km per second, and its diameter is 2 light-years.

■ Photographed by William Miller on Super Anscochrome 100 (90-minute exposure) with the 508-cm telescope on Mount Palomar.

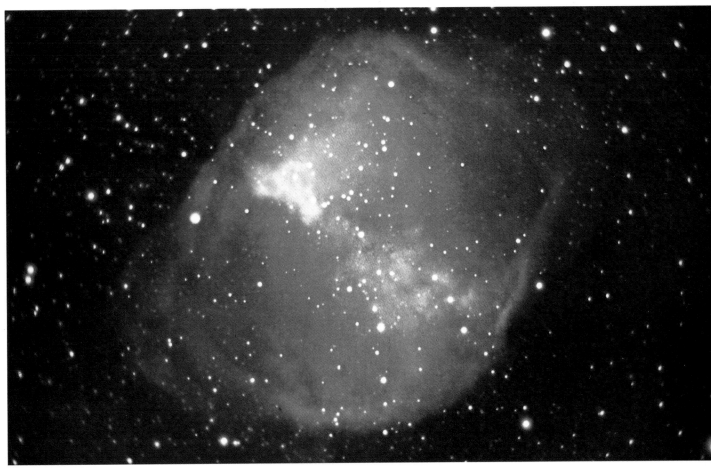

*The Dumbbell Nebula (M27) is the easiest planetary nebula to see, even with small instruments; it is visible even with binoculars. It was discovered in 1764 by Charles Messier, the great French comet hunter who compiled the first catalog of diffuse and nebular objects so as not to become confused in his research into objects (comets) that were similar in appearance to these. Messier's catalog contains more than 100 star clusters, galaxies and nebulas designated with an initial M and a progressive number. This is how he described M27: "It can be clearly seen through a telescope with a focal length of three and a half feet... it appears oval and does not contain stars. I will register it on the comet map of 1779." William Herschel viewed it as "a double layer of wideranging stars." The name, inspired by gym equipment, was coined by Reverend Thomas William Webb in the 19th century.*

Right, massive stars, after burning carbon and oxygen in their core at about 1 billion degrees C and silicon at 5 billion degrees C, retain a core of iron, surrounded by layers which, decreasing in temperature and density from the inside out, burn off less complex elements, including sulfur, chlorine, carbon, neon, helium and hydrogen. This "onion-like" stellar structure precedes a Type-II supernova colossal explosion, in which the star scatters much of its mass into space (arrows pointing outward), while the rest implodes, forming a neutron star (arrows pointing inward).

In the pictures at right and above right, the region of the Large Magellanic Cloud near the Tarantula Nebula, the scene of supernova 1987A's explosion, the closest supernova to Earth in recent years and one of the most important recent events in the field of astronomy. Above right, before the event; at right, 4 days after the explosion, which occurred on February 23, 1987.

■ Both images were produced from plates exposed in blue light (60 minutes on Kodak IIa-O with a GG 385 filter), green light (40 minutes on 103a-D with a GG 495 filter) and red light (60 minutes on 098-04 with an RG 630 filter), using the ESO's 1-m Schmidt telescope.

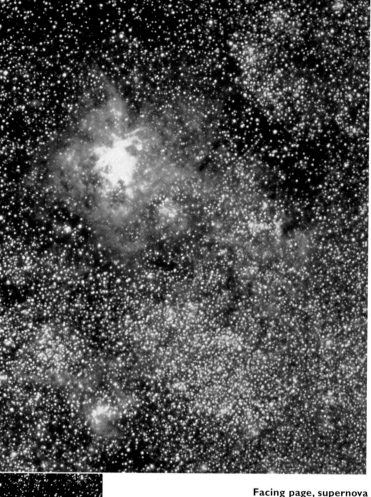

Facing page, supernova 1987A photographed in March 1987. The star which exploded, Sanduleak-69°202, was a blue supergiant, which increased in brightness during the explosion by about 4,000 times.
■ Tricolor image created by David Malin from three plates exposed on the **AAT** by Ray Sharples for 20, 15 and 20 minutes.

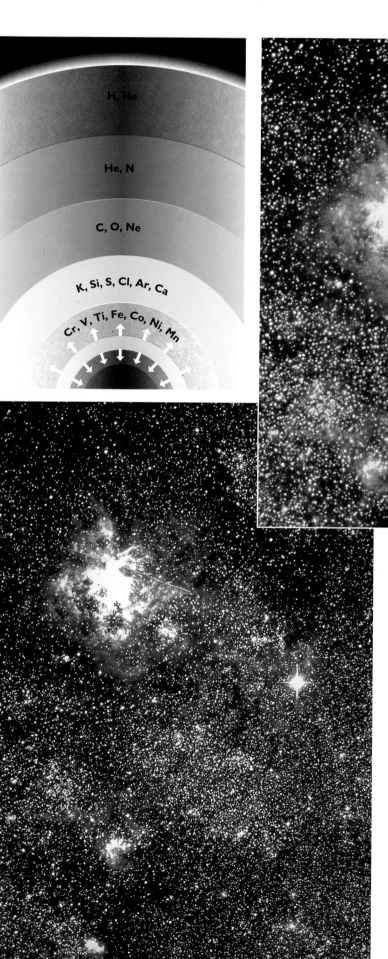

In 1987, the appearance of supernova 1987A raised a few questions. Until then, it had been believed that the precursor stars of Type-II supernovas were red supergiants. However, when it was discovered (by examining a plate that had been taken before the event) that it was a blue supergiant which had exploded, some believed that the entire theory about the last stages in the life of massive stars warranted a complete review. The spec-

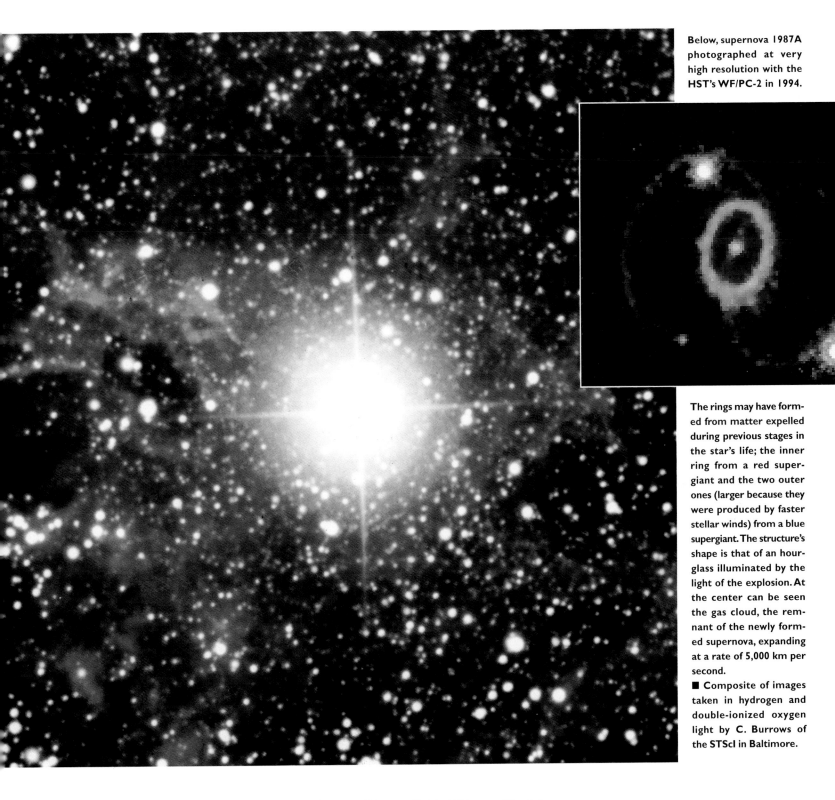

The rings may have formed from matter expelled during previous stages in the star's life; the inner ring from a red supergiant and the two outer ones (larger because they were produced by faster stellar winds) from a blue supergiant. The structure's shape is that of an hourglass illuminated by the light of the explosion. At the center can be seen the gas cloud, the remnant of the newly formed supernova, expanding at a rate of 5,000 km per second.
■ Composite of images taken in hydrogen and double-ionized oxygen light by C. Burrows of the STScI in Baltimore.

ulation was that the precursor of supernova 1987A had a peculiar chemical composition, exceptionally deficient in heavier elements, especially oxygen (typical of the Large Magellanic Cloud in which it was situated). Oxygen plays a particular role in inhibiting stellar contraction by favoring the production of energy through hydrogen fusion and by making the outer layer of a star's atmosphere more opaque to radiation. Therefore, a pauci-ty of oxygen could have modified the initial stages of the star's evolution, favoring the creation, before the explosion, of a blue supergiant. An alternative theory suggests that the parent star had really been a double star, in which the transference of mass from one star to the other could have favored the transformation of the primary star into a blue supergiant.

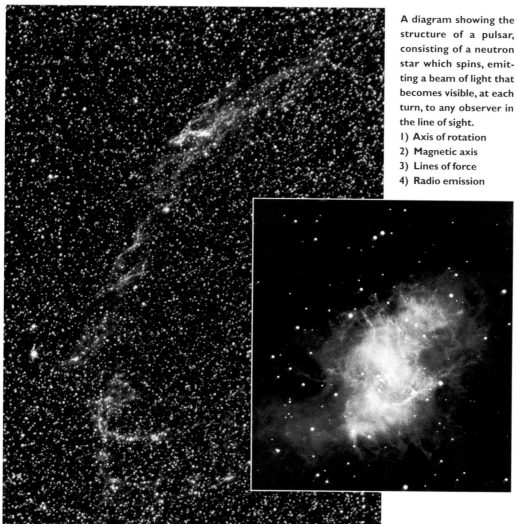

A diagram showing the structure of a pulsar, consisting of a neutron star which spins, emitting a beam of light that becomes visible, at each turn, to any observer in the line of sight.
1) Axis of rotation
2) Magnetic axis
3) Lines of force
4) Radio emission

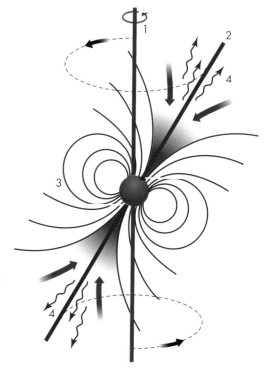

Facing page, the remnant of a supernova that exploded in the constellation Vela about 10,000 years ago. Since then, the gas has spread out over an area that is about three times larger than the Veil Nebula in Cygnus.

■ Tricolor composite by David Malin, using plates exposed at the UKS by Malcolm Hartley and Peter Standen for 60 minutes.

Above, NGC 6992, part of the Veil Nebula in Cygnus, the remnant of a supernova as large as six times the apparent diameter of the full Moon.

Center, the Crab Nebula, the remnant of the explosion of the supernova of 1054. A famous pulsar is located at the center of the nebula.
■ Both pictures were produced by William Miller at Mount Palomar on Super Anscochrome 100: NGC 6992 with the 122-cm Schmidt (a 2-hour exposure), the Crab with the 508-cm telescope (a 4-hour exposure).

The table at right gives relevant data for supernovas in our own galaxy historically documented up to the present day. Three events, however, remain questionable: that of 185 A.D. could refer to a comet, those of 386 and 393 to a normal nova. Of course, in order better to study the supernova phenomenon, it would be desirable to verify another close instance of this as soon as possible. Strangely, though, since the invention of the telescope, no supernova has appeared in our own galaxy.

| DATE (A.D.) | CONSTELL. | MAGN. | PLACE OF OBSERVATION | DURATION | DISTANCE (LIGHT-YEARS) | REMNANT |
|---|---|---|---|---|---|---|
| 185 | Centaurus | -3 | China | 7 months | 14 000 | gaseous optical and X-ray remnant (MSH 15-52); radio and X-ray pulsar (PSR 1509-58) |
| 386 | Sagittarius | 0 | China | 3 months | 16 300 | gaseous remnant (G11.2-0.3) |
| 393 | Scorpius | 0 | China | 8 months | 15 000 | unidentified |
| 1006 | Lupus | -9 | China/Japan Korea/Europe Arab countries | 24 months | 4 500 | gaseous optical and radio remnant (PKS 1459-41) |
| 1054 | Taurus | -5 | China/Japan Arab countries | 22 months | 6 500 | gaseous optical and X-ray remnant (Crab Nebula); Crab pulsar |
| 1181 | Cassiopeia | 0 | China/Japan | 6 months | 8 500 | radio source (3C 58) |
| 1572 | Cassiopeia | -4 | China/Korea Europe | 16 months | 7 500 | gaseous remnant and radio and X-ray source (Tycho's SNR) |
| 1604 | Ophiuchus | -3 | China/Korea Europe | 12 months | 14 300 | gaseous remnant and radio source (Kepler's SNR) |

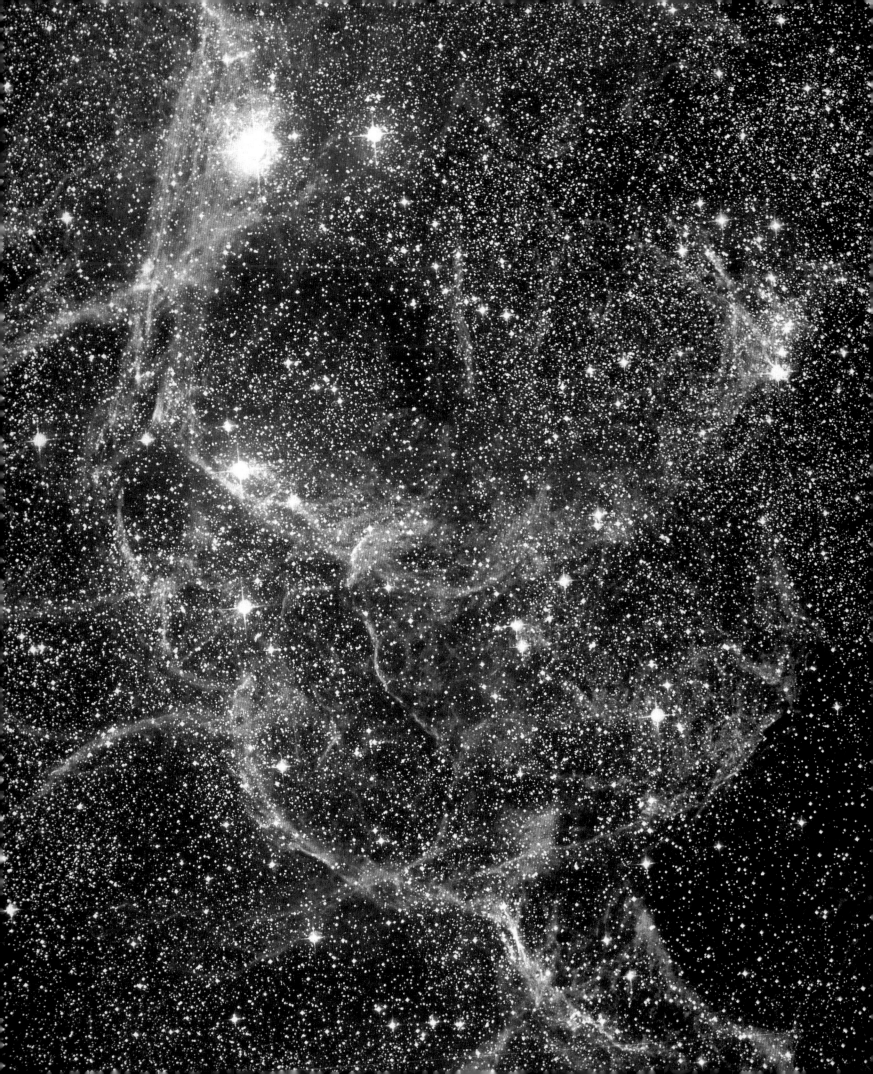

# Star clusters.

**Below, the faint nebulo-sity around the stars of NGC 3293, a typical open cluster, is a remnant of the gas cloud from which the stars formed.**
■ **Tricolor image created by Malin from three plates exposed on the AAT for 5 minutes each.**

As we have already seen, stars tend to evolve together, forming clusters of tens, hundreds, thousands, in some cases even millions of stars. Some clusters have been known since antiquity, such as the Coma Berenices star cluster, the Pleiades and the Hyades (the latter two belonging to the constellation Tau-rus). The Beehive (also known as Prae-sepe), situated in the constellation Cancer, can be seen with the naked eye as a fuzzy patch. In the third century B.C., the Greek poet Aratus, author of a famous description of the starry sky, *Phenomena*, referred to it as "a small haze," while the astronomer Hip-parchus, a century later, called it "a lit-tle cloud." Even the Double Cluster in Perseus was known since antiquity, though it was considered to be a patch of haze by both Hipparchus and Ptole-my (second century A.D.).

All of the above are open clusters, groups in which the stars are not very close to one another: their average dis-tance from each other is on the order of half of the average distance between the nearest stars surrounding our Sun—about 2 light-years. These groups are concentrated along the plane of our galactic disk (see figure below), the great spiral-shaped star city in which we live (see page 80). The pressure of gas within the spiral arms continually generates new stars from the nebulas scattered along the arms (see page 88). In theory, therefore, open clusters can be any age, because the star formation in a galaxy such as our own is still a long way from being complete. In reali-ty, however, after some time has elapsed following their formation, the clusters tend to break up, because each star follows its own orbital path around the center of the galaxy. Con-sequently, we currently can observe only recently formed open clusters, those in which this separation of the components has not yet occurred (apart from unusual cases with very numerous stars or stars that are ex-ceptionally close to one another).

The distances at which the open clus-ters lie range from a few dozen light-years up to tens of thousands (in the event that there are any on the other side of the galaxy). However, because they are concentrated, as we have seen, along the galactic plane, the light of the more distant ones is hidden by the stars, dust and gas in between.

There is a second category of clusters, the globulars, so called because the stars in them are so closely packed that they give the cluster the overall appearance of a globe-shaped struc-ture. Stars in these clusters are much more numerous, from tens of thou-sands to as many as millions, and they are closer to one another, as little as light-days or even light-hours apart. Globular clusters are spread out within the galactic halo, the roughly spherical formation that surrounds the disk and in which even single stars are im-mersed.

The size of globular clusters is on aver-age about double that of open clusters, but their current distribution and posi-tion within the halo ensures that they

**Right, a model of the Milky Way Galaxy show-ing its main compo-nents and the locations of open and globular clusters.**
**1) Position of our Sun**
**2) Nucleus**
**3) Nuclear bulge**
**4) Galactic disk**
**5) Globular clusters**
**6) Open clusters**

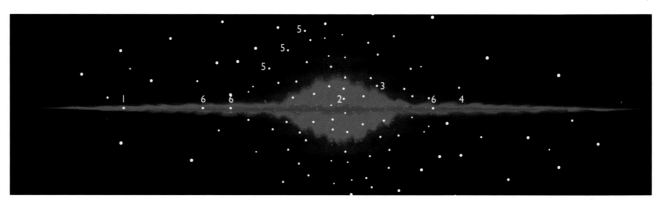

are all sufficiently distant not to be easily seen by the naked eye. Indeed, the ancients knew only Omega Centauri, which was cataloged by Ptolemy as a fuzzy star.

In globular clusters, the process of star formation occurred practically in a single moment, at the time the Milky Way was formed. Globular clusters represent the first stage in the contraction of great clouds of gas that eventually produced considerable numbers of stars. Immediately afterward, the protogalaxy, the large cloud from which our solar system evolved, rapidly collapsed, forming the disk. This left relatively devoid of gas those regions of the halo that were subsequently emptied by the furious stellar winds originating from the many supernova explosions which ended the life of the more massive stars that had formed in the meantime. All this blocked any further stellar synthesis.

Globular clusters are nevertheless the oldest objects in the galaxy. Their stars are about as old as the universe, at least 10 billion years. These are known as Population II stars. Open clusters, on the other hand, are made up of stars of all ages, but generally much younger ones. These are Population I stars, formed from the "ashes," rich in complex elements, ejected from Population II stars and scattered through space by the stellar winds and by nova and su-

pernova explosions.

The age of a cluster is determined by plotting a graph correlating the luminosity and the temperature of the stars in the cluster with those of other clusters (see figure at right). Most stars are arranged along the main sequence, which corresponds to their stable phase of life, in which they burn up the hydrogen contained in their core.

The main sequence appears to branch off lower down into old clusters, such as the globular cluster M3 or the open cluster M67, and, much higher up, into young clusters, such as the Pleiades or NGC 2362. This can be explained by considering that a star's life span is inversely proportional to the cube of its mass: a star of 10 solar masses lives only one-thousandth of the life span of the Sun. In young clusters, only the stars of greater mass have already ended their stable life, leaving the main sequence, while those of small or average mass are still present. In older clusters, on the other hand, even the smaller stars have left the main sequence. Diagrams of this type help to calculate the age of individual clusters, although the results that can be obtained from them

are, clearly, only approximate. It has thus been noted that while the age of open clusters ranges from 1 million to 10 billion years, that of globular clusters is generally close to the upper end of that range.

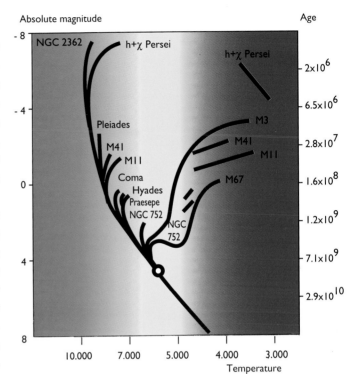

**Above, temperature-luminosity diagram for a few star clusters. This is known as a Hertz-sprung-Russell (H-R) diagram, after Danish astronomer Ejnar Hertzsprung and American astronomer Henry Norris Russell, who originated it between 1911 and 1913. Plotted on the vertical scale on the left side, the absolute magnitude is a measure of the luminosity that a celestial object would show if it were placed at a standard distance of 10 parsecs (1 parsec is equivalent to 3.26 light-years). On the right-hand vertical scale, the age in years is given. Temperature in absolute degrees (Kelvin) is plotted along the bottom. The empty circle represents the position of the Sun.**

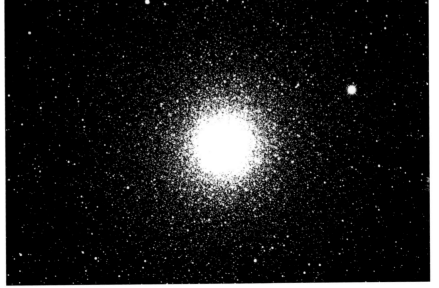

**A typical globular cluster, M15 in Pegasus, photographed with the 3.8-m Mayall Telescope of the KPNO. It is 31,000 light-years away, with a diameter of 110 light-years.**

The Pleiades, an open cluster 410 light-years away, containing hundreds of stars. The brightest are young giant blue stars with luminosity 100-1,000 times that of the Sun. They are concentrated in an area 7 light-years across, while the entire group covers an area 40 light-years across. The stars in the Pleiades are about 150 million years old. Of all celestial objects, the Pleiades has been most frequently observed and photographed or recorded by all generations of astronomers since time immemorial. The Pleiades are mentioned in the texts and oral legends of all civilizations in the world. Some people even regulated their civil life according to the seasonal positions of the Pleiades. All cultures have given them proper names: *the Seven Sisters, the Little Hens, the Little Eyes, the Bunch of Grapes, the Jewel Chest, the Old Wives, the Seven Goats, the Seven Doves, the Sailing Stars, the Chariot and the Company of Maidens* are just a few. The great poets have sung their praises, including Homer, Sappho, Aratus, Hesiod, Virgil, Ovid and Tennyson. The gas (nebulosity) visible in the photograph does not belong to the Pleiades but is present by chance in the part of space currently being traversed by the cluster. ■ Photographed by William Miller on Super Anscochrome 100; a 90-minute exposure, taken with the 122-cm Schmidt telescope on Mount Palomar.

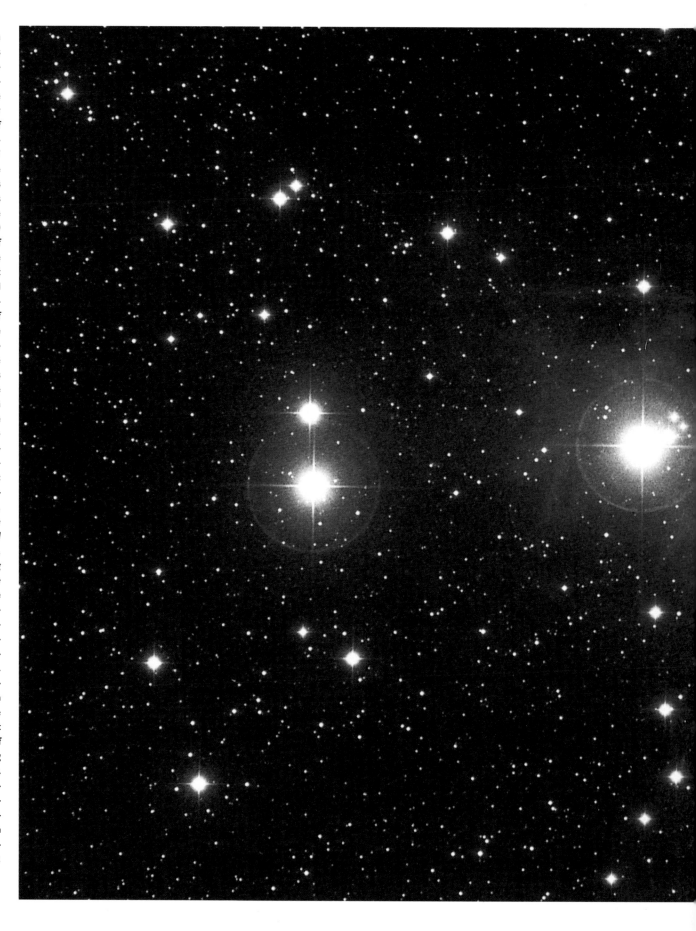

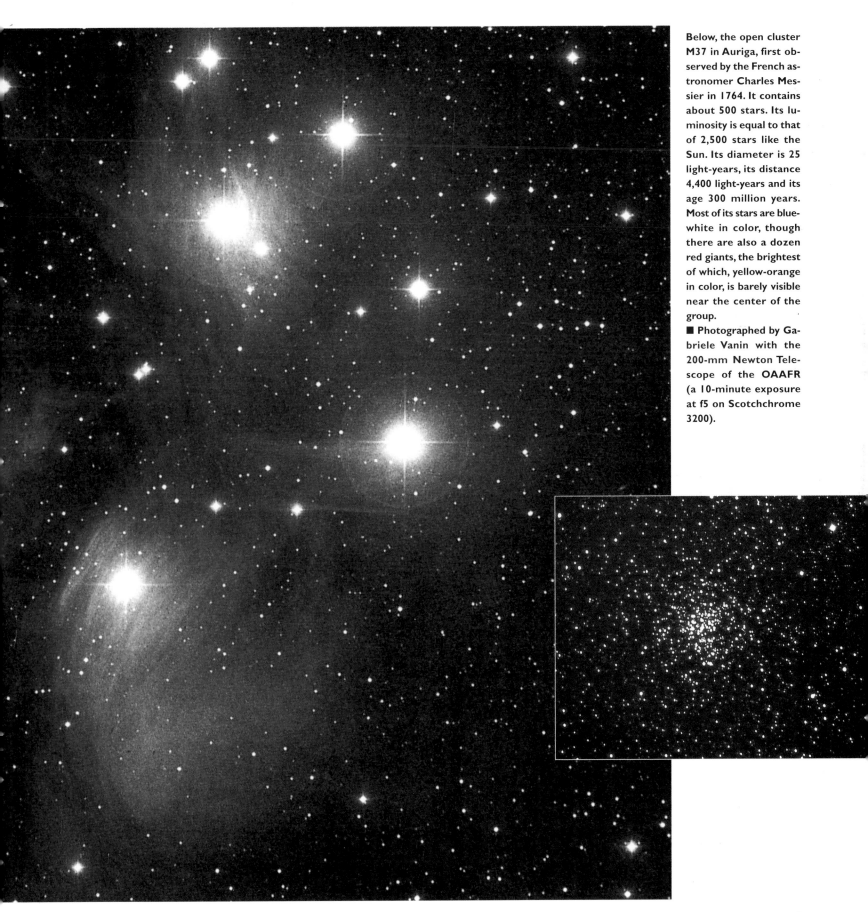

Below, the open cluster M37 in Auriga, first observed by the French astronomer Charles Messier in 1764. It contains about 500 stars. Its luminosity is equal to that of 2,500 stars like the Sun. Its diameter is 25 light-years, its distance 4,400 light-years and its age 300 million years. Most of its stars are blue-white in color, though there are also a dozen red giants, the brightest of which, yellow-orange in color, is barely visible near the center of the group.
■ Photographed by Gabriele Vanin with the 200-mm Newton Telescope of the **OAAFR** (a 10-minute exposure at f5 on Scotchchrome 3200).

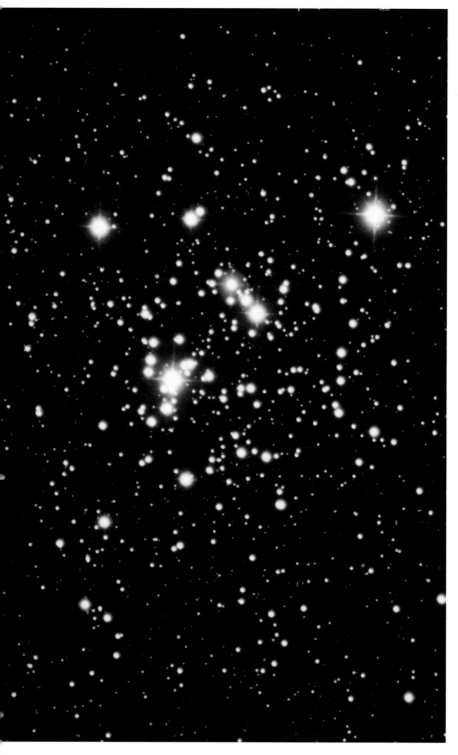

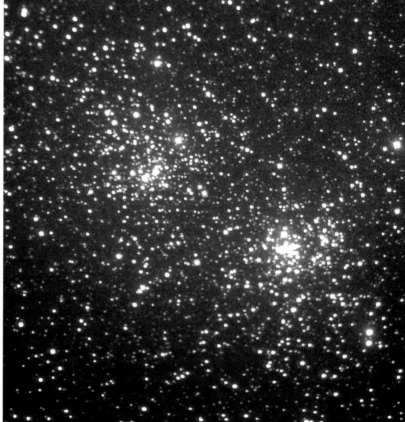

Left, the open cluster **NGC 4755**, the "Jewel Box," in Crux, the Southern Cross. It is 7,600 light-years away, and its diameter is 25 light-years. It is a fairly young group, just 7 million to 15 million years old. The brightest stars are highly luminous blue giants: the two most luminous ones have a brightness 80,000 times that of the Sun. The cluster's young age has not prevented one component from evolving toward the red supergiant stage (center). ■ Tricolor image created by David Malin from three plates exposed by him at the AAT for 5 minutes each.

Above, **NGC 869-884**, the Double Cluster in Perseus, containing about 700 stars in all. The 10 brightest are blue giants with a luminosity of 7,500 to 70,000 times that of the Sun. In the two groups, there are also about 20 red supergiants. Various studies have demonstrated not only that **NGC 869** and **NGC 884** are not physically connected but also that they are not even the same age and are not located at the same distance from us. **NGC 869** is nearer (7,200 light-years, as opposed to 7,500 for **NGC 884**). Both are extremely young, however, with an age that is probably no greater than 6 million years. Their combined diameter is 70 light-years. ■ Photographed by Gabriele Vanin with the 200-mm Newton Telescope of the OAAFR (a 30-minute exposure on Scotchchrome 1000 at f5).

*The poetic name for NGC 4755—the Jewel Box—was coined by John Herschel during his sojourn at the Cape of Good Hope from 1834 to 1838 to explore the southern skies. He described it thus: "The stars which make it up, seen through a telescope large enough to show the stars' colors, resemble a casket of variously colored precious stones." Unfortunately, this wonder can be viewed only from the southern hemisphere.*

*NGC 869-884, on the other hand, is a marvel reserved only to those who live in the northern hemisphere. Its existence has been known since at least 150 B.C., but its true nature was only understood with the invention of the telescope. It is also known as the "sword hilt of Perseus," from the constellation in which it is found, named after the mythological hero who slew Medusa. One of the mysteries surrounding NGC 869-884 is why French*

*astronomer Charles Messier never included it in his catalog of nebular bodies. The opinion that he failed to do so because the Double Cluster was so well known that he was in no danger of confusing it with a new comet does not hold up, because Messier cataloged other clusters that were as—if not more—famous, such as the Beehive (M44), in the constellation Cancer, and the Pleiades (M45).*

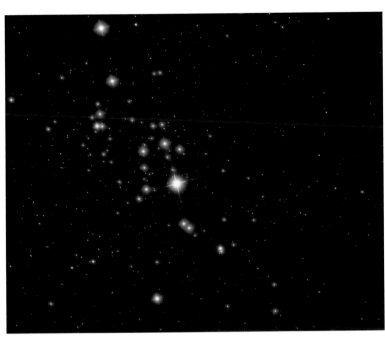

Below, the open clusters M36 (left) and M38 in Auriga are about 2.3 degrees apart and have apparent diameters of 12 and 21 arc-minutes, respectively, and real diameters of 14 and 25 light-years. Their respective distances from Earth are 4,100 and 4,300 light-years. M36 contains about 60 stars and is a very young cluster (25 million years). M38 contains about 100 stars and is 200 million years old.

■ Ten-minute exposure on Scotchchrome 3200 taken with the OAAFR's 140-mm Comet Catcher at f3.5 (diaphragmed to f5).

Above, the Alpha Persei cluster. At center of photo is the brightest star in the constellation Perseus, Mirfak (Alpha Persei), which belongs to one of the closest star clusters. At a distance of 550 light-years and having an apparent diameter of 3 degrees, the cluster spans almost 30 light-years of space. It contains about 100 stars, is 5,000 times brighter than the Sun and is estimated to be 50 million years old.

■ Photographed by Gabriele Vanin. Six-minute exposure on Scotchchrome 3200 with a 180-mm lens at f4, using diffusion and cross-screen filters.

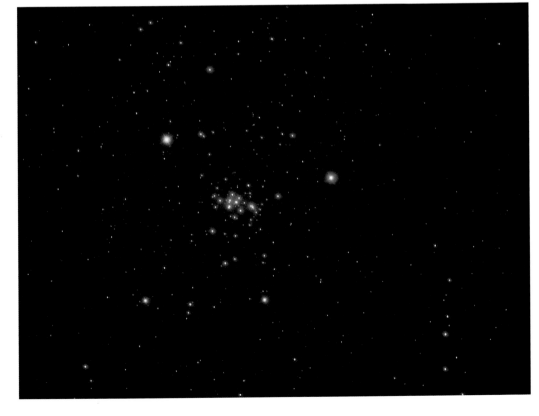

At left, M44, the Beehive cluster. This wide-field photo shows the cluster located in the middle of the square formed by the stars Gamma, Eta, Theta and Delta in the constellation Cancer. The Beehive contains about 200 stars and has an apparent diameter of 95 arc-minutes and a real one of 40 light-years. It is 520 light-years away and is estimated to be 660 million years old.

■ Photographed by Gabriele Vanin. Six-minute exposure on Scotchchrome 3200 taken with a 180-mm lens at f3.5 using cross-screen and diffusion filters.

*The Beehive cluster (M44), also known as Praesepe, is readily seen with the naked eye and therefore has been known since antiquity. Aratus wrote about it in the third century B.C., calling it a "little fog," and in the second century B.C., Hipparchus referred to it as a "little cloud." In ancient times, it was used as a weather warning sign—its invisibility in an otherwise calm sky was said to indicate the approach of a storm. Not until 1610 did Galileo reveal its true nature.*

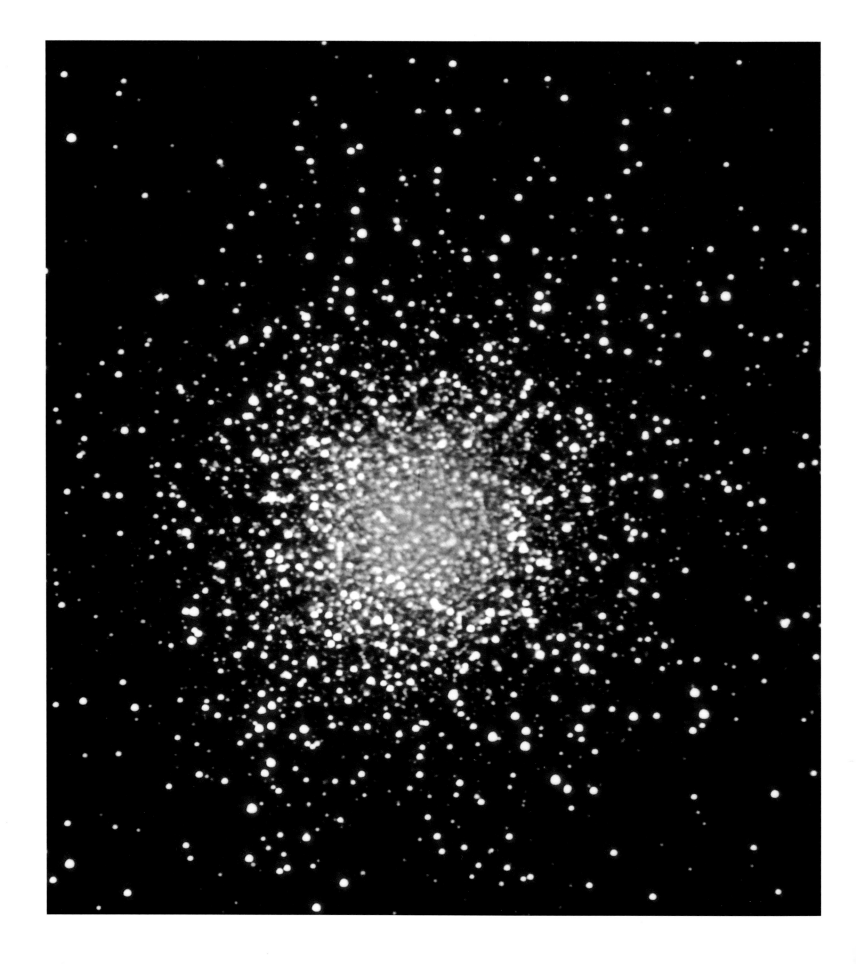

Facing page, M13 in Hercules, the brightest globular cluster visible from the northern hemisphere. Its distance is 23,500 light-years, and its diameter is 155 light-years. Its luminosity is equivalent to more than 200,000 times that of the Sun, and it contains no fewer than 1 million stars. The brightest are red giants, with a luminosity 1,500 times that of the Sun. Its age is estimated at 10 billion years.

■ Photograph taken with the 155-cm telescope at the US Naval Observatory station in Flagstaff, Arizona.

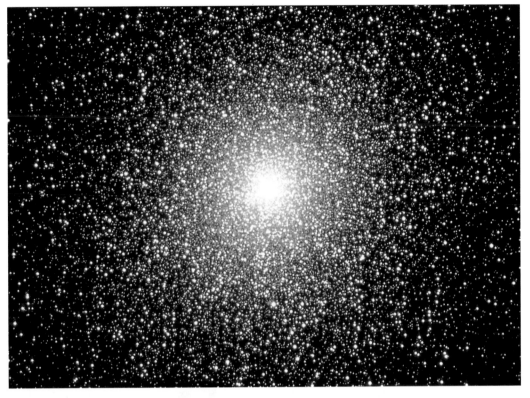

Left, 47 Tucanae, the second most luminous globular cluster in the sky, unfortunately also visible only from the southern hemisphere. It is 15,000 light-years away and has a diameter of 200 light-years. Its luminosity is equal to half a million times that of the Sun, and it contains at least 2 million stars. Compared with most of its peers, 47 Tucanae is a young globular cluster: it is about 7 billion years old.

■ Tricolor composite created from three plates exposed in blue, green and red light by G. Pizarro using the ESO's 1-m Schmidt telescope.

Right, Omega Centauri, the brightest globular cluster in the sky, is visible only from the southern hemisphere. It is 17,000 light-years away, and its diameter is 350 light-years. Recent estimates of its mass carried out by Georges Meylan at the ESO give it a value of 5 million solar masses. Omega Centauri, thus, could be considered halfway between an ordinary globular cluster and a dwarf elliptical galaxy (see page 85). Its age is similar to that of M13.

■ Picture taken with the CTIO's 4-m telescope.

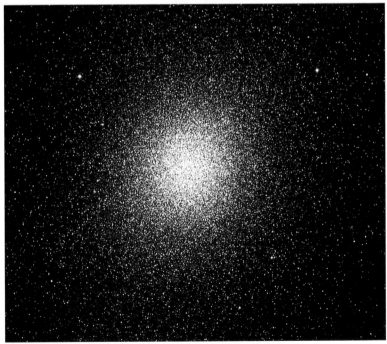

Above, Omega Centauri photographed in ultraviolet light, at 160 nm, with the 38-cm Ultraviolet Imaging Telescope on NASA's Astro satellite, which flew on-board the shuttle Columbia in December 1990. At this wavelength, the hottest stars in the cluster are visible.

*Omega Centauri, known to the ancients as a star, was cataloged as such even by Bayer, inventor of the modern method of naming stars by assigning a Greek letter in order of luminosity and/or position in a constellation, followed by the possessive form of the constellation's Latin name, a convention introduced in his atlas Uranometria of 1603. English astronomer Halley first resolved the cluster into separate stars in 1677. In 1834, John Herschel*

*noted that it was "... without compare, the largest and richest body of its kind in all the skies."*
*47 Tucanae is easily visible to the naked eye, next to the Small Magellanic Cloud (the very bright "star" just right of the Cloud in the photo, page 84). Naturalist von Humboldt, observing it from Peru in the early 19th century, mistook it for a comet. As far back as 1775, however, Abbé de Lacaille (page 102) discovered its true nature.*

*M13 can be just glimpsed with the naked eye on very clear nights, as Halley did, who discovered it in 1714. In 1764, it appeared to Messier as a nebula, since he was unable to resolve even one star, and this tells us much about the optical quality of the telescopes of the time. It was William Herschel who first discovered M13 to be a cluster, estimating that it contained 14,000 stars.*

Facing page, M104, the Sombrero Galaxy in Virgo, 40 million light-years away. This is a spiral galaxy with an enormous central bulge. In the halo, a few dozen globular clusters may be seen.

■ Tricolor composite produced from plates taken with the ESO's 3.6-m telescope. Globular clusters NGC 1818 (at right) and NGC 1850 (below left) in the Large Magellanic Cloud photographed with the Space Telescope's WF/PC-2. Below, the center of 47 Tucanae photographed in ultraviolet light with the FOC before and after the application of COSTAR (page 123). The resolution is suf-

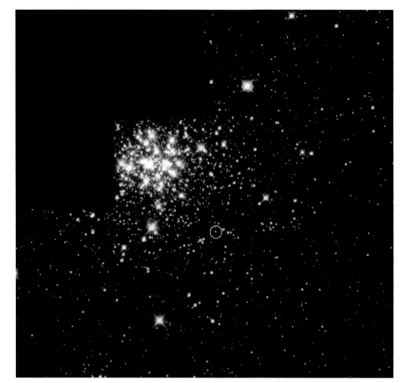

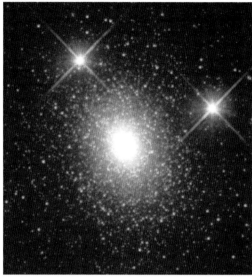

Above, a Hubble Space Telescope image of the globular cluster G1 in the halo of the Andromeda Galaxy. Hubble's view of this distant globular is much like the appearance of the Milky Way's globular clusters as seen in ground-based telescopes, even though G1 is 100 times farther away.

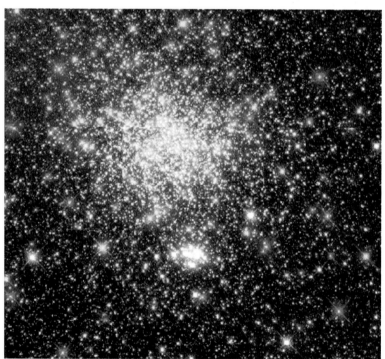

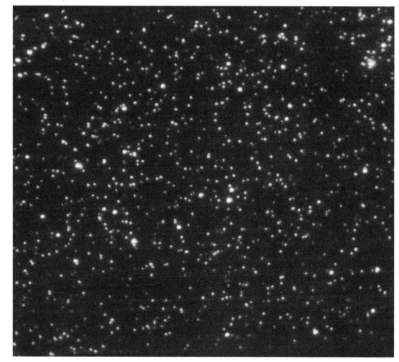

ficiently improved to show stars 15 times fainter and separated from one another by only 30 light-years. Pictures such as these enable us not only to exclude the existence of black holes at the center of globular clusters but also to detect white dwarfs, which is essential for reconstructing the dynamic and evolutionary history of the clusters.
■ Pictures taken by R. Jedrzejewski of NASA and by F. Paresce and G. De Marchi of the ESA.

*Since the repairs to the Hubble Space Telescope, it is now possible to address one of the most crucial problems in the dynamics of globular clusters. Various studies carried out in recent years estimated that the stars in the clusters' nuclei must collapse, forming a black hole, an object so compact that the gravity at its surface will not allow even light to escape. However, observations carried out so far with the Hubble on M14, M15 and 47 Tucanae reveal a different picture, according to which the collapse is avoided because groups of contact binary stars are created at the heart of the clusters. The rapid orbital motions of these stars act as energy reservoirs upon which other stars can draw, increasing their velocity and making the structure expand. It is probable that the clusters live complete cycles of expansion and contraction.*

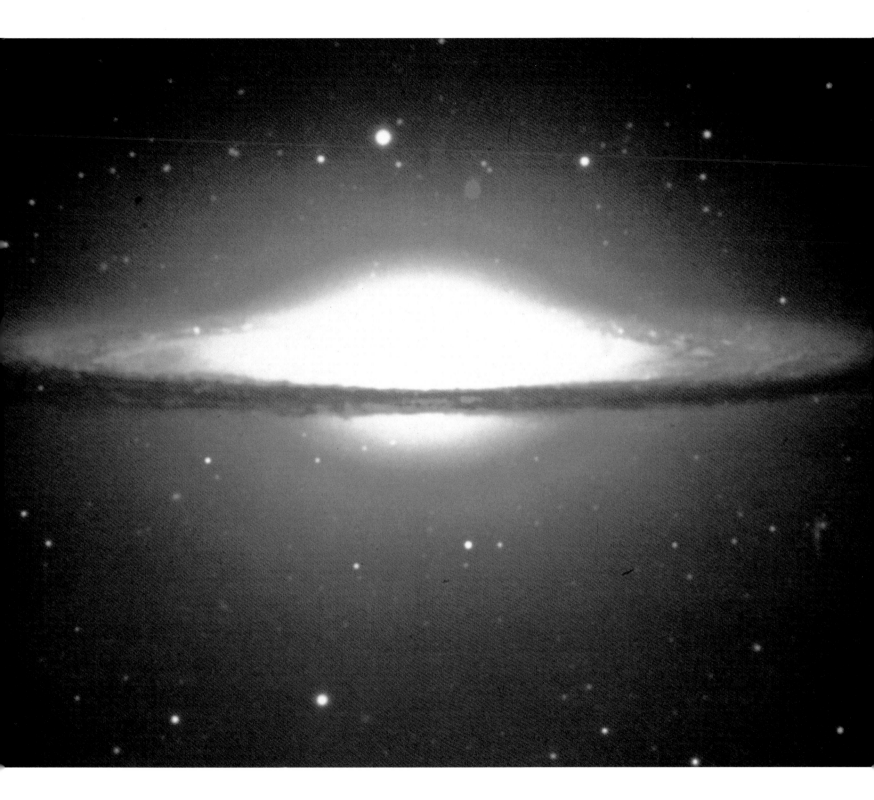

The study of the globular clusters of other galaxies is very important not only to understand the origin, evolution and dynamic behavior of galaxies (including our own) but also to refine our ideas about stellar evolution. Only 150 globular clusters belonging to the Milky Way Galaxy are known, a sample that is not very significant for statistical studies of the life of stars. On the other hand, the globular clusters of other galaxies are so distant (over 15 million light-years) that only large telescopes in very favorable locations are able to glimpse them (let alone study them in detail). M87, for example (see page 97), contains hundreds of globular clusters. In recent years at Mauna Kea, Hawaii, several astronomers have succeeded in gathering excellent pictures showing globular clusters around other galaxies, using CCD cameras fitted onto the 3.6-m Canada-France-Hawaii Telescope.

In 1987, William Harris, Laird Thompson and Frank Valdes succeeded in resolving globular clusters in NGC 4874 and NGC 4889, in the Coma galaxy cluster, 270 million light-years away. In 1990, William Harris and Christopher Pritchet photographed globulars in NGC 6166, a member of the Abell 2199 cluster, 375 million light-years away.

# From the Galaxies to the Universe.

All the stars we can see in the sky on a clear night, as well as the star clusters and nebulas, are part of a single large system known as the Milky Way Galaxy.

As far back as the mid-16th century, scientists began to realize that the universe was not confined just to the solar system. The English Copernican Thomas Digges suggested in 1576 that the stars were scattered in space at different distances from us, rather than being fixed on a single sphere just outside the orbit of the farthest known planet (as even Copernicus had maintained). This idea was taken up at the end of the 16th century from a philosophical angle by Giordano Bruno, who stated that the "fixed stars" were objects like our Sun, other suns around which planets similar to Earth and the others in our system could orbit. Telescopic observations carried out by Galileo confirmed this assertion. Looking through a telescope at the Milky Way, that milky strip which crosses the sky especially on summer nights (see photographs on this page and opposite), Galileo was able to resolve it into stars. The stars appeared to be scattered at all distances from us, some so far away that the naked eye could not even detect them. In the 17th century, therefore, an idea of the vastness of this star system became clear, and it was considered to be the entire existing universe.

It was William Herschel who first tried to understand the span and structure of "the Galaxy," a term which is sometimes used for the Milky Way (*gala* is Greek for "milk"). Herschel adopted the method of counting with a telescope the stars present in selected areas of the sky in order to get an idea of their distribution in space. He concluded that the Galaxy had a decidedly flat and highly irregular shape, with the Sun occupying its center (see figure below). In this way, the sky's appearance could easily be explained: looking along the plane of the Galaxy, it is possible to see many closely packed stars, forming a diffuse and indistinct light, namely the Milky Way, whereas in directions perpendicular to the plane, far fewer stars are seen.

No substantial progress was made in terms of understanding the dimensions of the Milky Way Galaxy until 1838, when the first measurements of stellar distances were achieved and it was realized that the star system must be at least a million times wider than the solar system. Until about 1910, scientists thought the Sun was at the center of the Milky Way Galaxy and that this was the entire universe. At that time, Dutch as-

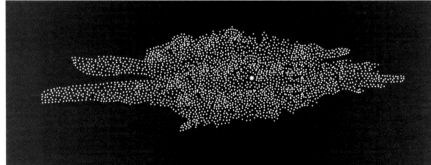

■ **Photograph taken with a 35-mm lens at f2.8 with an exposure time of 5 minutes on Scotchchrome 3200.**

**Above, model of the Galaxy constructed by William Herschel on the ba-** sis of star counts carried out by him and presented to the Royal Society in 1785 in a lecture titled "On the Structure of the Heavens." The largest dot represents the Sun, which, according to Herschel, was situated at the center of a "truly wide, ramified and composite mass of many millions of stars."

tronomer Jacobus Kapteyn, using star counts much more detailed than Herschel's, assigned the Galaxy the shape of a somewhat flattened ellipsoid with a diameter of about 30,000 light-years. Kapteyn was, however, still convinced that the Sun was at the center of the Galaxy.

Around 1920, American astronomer Harlow Shapley demonstrated that the distribution of globular clusters (see figure on page 70) was eccentric in relation to the Sun and that therefore (assuming for simplicity's sake that the clusters themselves were arranged around the galactic center), the Sun was at the edge of the stellar system. Today, we know that the shape of the Milky Way Galaxy is a kind of pinwheel when viewed from the top (see model below) and a lenticular disk with a central bulge when viewed edge-on (see figure on page 70). Even Shapley, however, was convinced that the Galaxy was the entire universe. From measurements of the distances of globular clusters, he arrived at a diameter for the star system of about 300,000 light-years. It had been suspected for some time that some types of nebula visible through the telescope, the so-called "spiral nebulas," were much farther away, completely outside the Galaxy, and that therefore the universe was much wider than had been imagined. Herschel himself was of this opinion.

In 1920, a famous debate took place between Harlow Shapley and Heber Curtis, who held this opinion. Curtis won "on points," as we would say nowadays. Even without his needing to deliver a knockout punch, Curtis's idea that the universe stretched vastly beyond our own galaxy began to be shared by his astronomer colleagues.

In 1924-25, Edwin Hubble's research on the nearby galaxies M31, M33 and NGC 6822 demonstrated irrefutably that they were at least three times farther away than the maximum value assigned to the diameter of the Milky Way Galaxy by Shapley. Between 1922 and 1936, Hubble suggested a classification system for galaxies, dividing them into three categories according to their shapes: elliptical, spiral and irregular. Elliptical galaxies are further divided into subtypes according to their degree of flattening, from E0 for spherical ones to E7 for the flattest ones. Spiral galaxies are categorized as normal or barred. Irregular galaxies are those lacking any recognizable structure. Hubble also identified a type intermediate between spiral and elliptical galaxies, the S0 galaxies, with a disk structure but without spiral arms. During these same years, the great astronomer began to lay the foundation for modern cosmology and for the discovery of a rapidly ex-

panding universe, rich in innumerable galaxies situated at distances that would at one time have been considered absurd—up to billions of light-years.

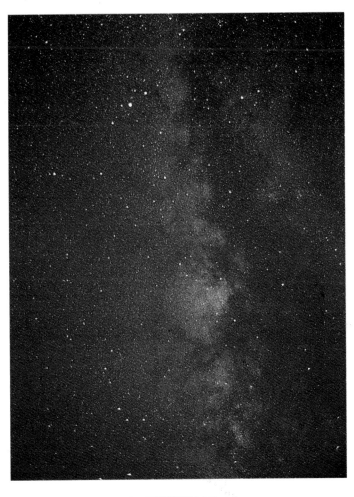

**Right, model of the Milky Way based on the most recent findings, according to which it is midway between a normal spiral and a barred spiral; that is, its nuclear bulge is elongated into a kind of bar.**

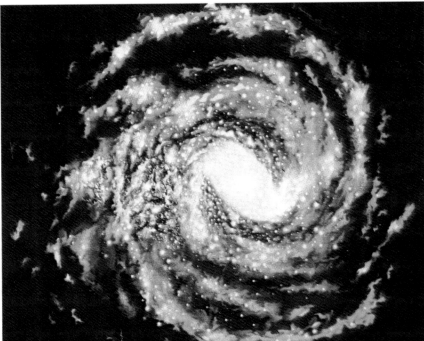

**Above, the portion of the Milky Way in the constellations Scutum and Aquila, photographed by Gabriele Vanin from the Rifugio Lagazuoi in the Dolomites, Italy. Although still clearly visible, the Milky Way is not as dense here as in the picture on page 80, because this sector is farther away from the galactic center.**
■ **Five-minute exposure with a 35-mm lens at f2.8 on Scotchchrome 3200 film.**

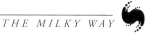

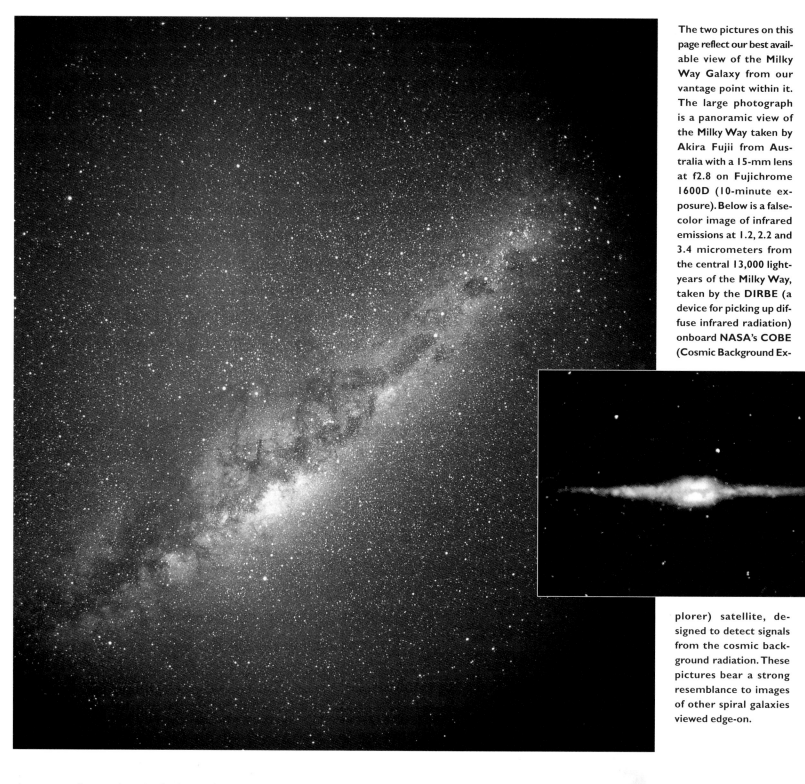

The two pictures on this page reflect our best available view of the Milky Way Galaxy from our vantage point within it. The large photograph is a panoramic view of the Milky Way taken by Akira Fujii from Australia with a 15-mm lens at f2.8 on Fujichrome 1600D (10-minute exposure). Below is a false-color image of infrared emissions at 1.2, 2.2 and 3.4 micrometers from the central 13,000 light-years of the Milky Way, taken by the DIRBE (a device for picking up diffuse infrared radiation) onboard NASA's COBE (Cosmic Background Explorer) satellite, designed to detect signals from the cosmic background radiation. These pictures bear a strong resemblance to images of other spiral galaxies viewed edge-on.

It was not at all easy to determine the shape and structure of our galaxy, a feat rather like trying to draw an overall map of an entire city while standing at one spot inside it (without being able to do an aerial reconnaissance). An important question concerned the spiral arms. Because these are always characterized by intense stellar birth activity, it is enough to look for traces of this continual star generation. Such traces are hot, young stars (such as Wolf-Rayet stars, with surface temperatures of 50,000 degrees C, and long-period Cepheid variables); areas of ionized hydrogen, resulting from the ultraviolet radiation produced by these stars; or very young open clusters. Through these indicators, it has been possible to deduce the structure of the spiral arms nearest to our Sun. For those that are farther away, clues are provided by the radiation emitted at a wavelength of 21 cm by the neutral hydrogen that is their main constituent, which can be picked up by radio telescopes. If we know the position of the radiation source and the speed with which it is moving away from us, its distance can be calculated by applying the appropriate mathematical formulas.

The center of the Milky Way Galaxy is the seat of intense energy phenomena, but it is completely opaque to visible radiation because of the powerful absorptive capacity of the interstellar dust between our solar system and the galactic nuclear bulge. (This is responsible for the characteristic reddish color of the galactic disk seen in the COBE picture on facing page.) Nevertheless, it is certainly possible to investigate the galactic nucleus by using those types of radiation with wavelengths larger than the dimensions of the interstellar dust particles: namely, infrared radiation and radio waves. On this page, we see two radio images obtained by the Very Large Array at Socorro, New Mexico, a radio telescope made up of a row of 27 antennas operating simultaneously. Below, an area 230 light-years across is visible around the galactic center. On the right is an area of strong radio emission, known as radio source Sagittarius A (Sgr A). At its center, the yellow dot represents Sgr A*, a compact source of radio waves that marks the position of the galactic nucleus. On the left, the mass of filaments corresponds to the energy emission of electrons moving in a spiral around the lines of force of a powerful magnetic field.

■ Image taken by F. Yusef-Zadeh, M.R. Morris and D.R. Chance at a wavelength of 20 cm with a resolution of 5 x 8 arc-seconds.

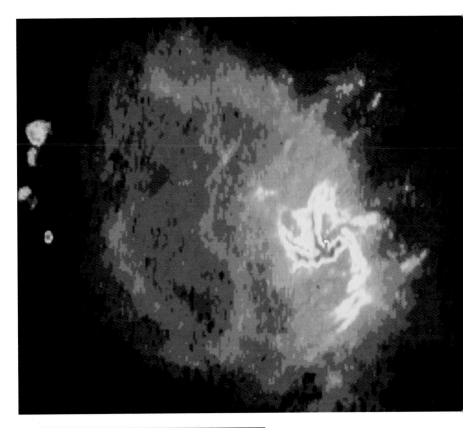

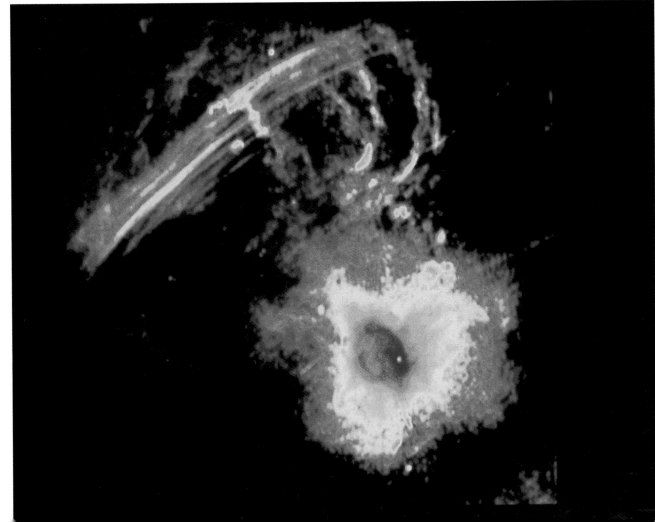

Above, a more detailed image of Sgr A resolves it into two components: Sgr A East (left), with energetic emissions derived from a long-ago supernova explosion; and Sgr A West, a spiral structure formed by ionized gases that appear to be falling in toward the galactic center (marked by the radio source Sgr A*, which is indicated by the area of reddest coloration). Nearby is a star cluster with a population density of 100,000 stars per cubic light-year. Hidden within one of these two structures, there might lie an enormous black hole, marking the true center of the Milky Way Galaxy.

■ Image taken by W.M. Goss, R.D. Ekers, J.H. van Gorkom and U.J. Schwarz at wavelengths of 20 cm and 6 cm and a resolution of 1.3 x 2.5 arc-seconds (about 45 light-years).

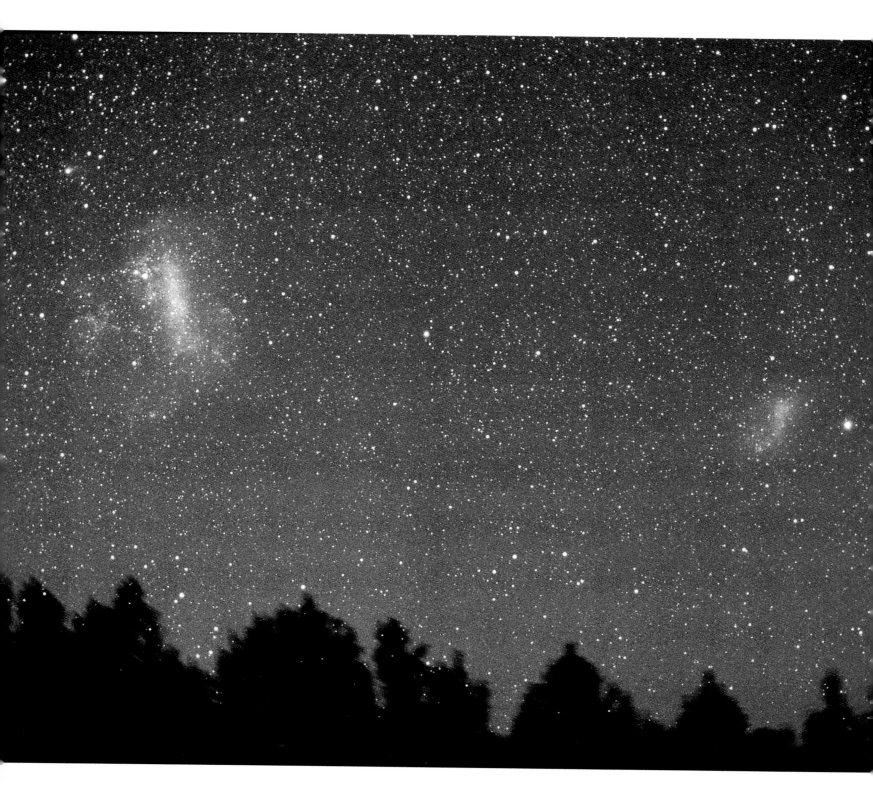

The two Magellanic Clouds, small irregular galaxies, are satellites of the Milky Way, together with about 10 other dwarf galaxies (pages 98-99). The Small Magellanic Cloud is on the right.
■ Photographed by A. Fujii from Australia, with a 35-mm lens and 10 minutes' exposure on Fujichrome 1600D.

*The Magellanic Clouds enjoy a particular importance in the history of astronomy for various reasons. One of the most significant is a major milestone in measuring extragalactic distances: it was by comparing a large number of photographs of the Clouds that Henrietta Leavitt discovered the period-luminosity relationship of the Cepheids (see page 47). It was completely reasonable to assume that the Large and the Small Cloud were true star systems and that, therefore, all the Cepheids that were visible in either of them must be at about the same distance from us. Thus, observed differences in their apparent luminosities must indicate actual differences in intrinsic luminosity, rather than an effect of distance.*

| NAME | TYPE | MAGNITUDE (IN BLUE) | DISTANCE (LIGHT-YEARS) | DIAMETER (LIGHT-YEARS) |
|---|---|---|---|---|
| Milky Way | S(B)bc | | | 100 000 |
| *Dwarf galaxy in Sagittarius | dSph | 3.0 | 50 000 | 10 000 |
| *Large Magellanic Cloud (LMC) | Irr | 0.6 | 170 000 | 20 000 |
| *Small Magellanic Cloud (SMC) | Irr | 2.8 | 200 000 | 15 000 |
| *Dwarf galaxy in Ursa Minor | dSph | 12.4 | 210 000 | 1 000 |
| *Dwarf galaxy in Draco | dSph | 11.9 | 240 000 | 500 |
| *Dwarf galaxy in Sculptor | dSph | 9.1 | 260 000 | 1 000 |
| *Sextans I | E | ? | 300 000 | 3 000 |
| *Dwarf galaxy in Carina | dSph | 10.6 | 330 000 | 500 |
| *Dwarf galaxy in Fornax | dSph | 8.5 | 450 000 | 3 000 |
| *Leo I | dSph | 11.8 | 720 000 | 1 000 |
| *Leo II | dSph | 12.3 | 750 000 | 500 |
| Barnard's Galaxy (NGC 6822) | Irr | 9.3 | 1 700 000 | 8 000 |
| IC 5152 | Irr | 11.7 | 2 000 000 | 5 000 |
| WLM System | Irr | 11.3 | 2 000 000 | 7 000 |
| Andromeda Galaxy (M31) | Sb | 4.4 | 2 200 000 | 125 000 |
| #M32 | E2 | 9.0 | 2 200 000 | 6 000 |
| #NGC 205 | E5 | 8.6 | 2 200 000 | 10 000 |
| #NGC 185 | E3 | 10.1 | 2 200 000 | 6 000 |
| #NGC 147 | E5 | 10.4 | 2 200 000 | 10 000 |
| #And I | dSph | 13.9 | 2 200 000 | 2 000 |
| #And II | dSph | 13.5 | 2 200 000 | 2 000 |
| #And III | dSph | 13.5 | 2 200 000 | 6 000 |
| Triangulum Galaxy (M33) | Sc | 6.3 | 2 500 000 | 45 000 |
| IC 1613 | Irr | 10.0 | 2 500 000 | 12 000 |
| LGS 3 in Pisces | Irr | 15.5 | 3 000 000 | 500 |
| Dwarf galaxy in Aquarius | Irr | 15.3 | 3 000 000 | 4 000 |
| Dwarf galaxy in Tucana | dSph | 15.3 | 3 000 000 | 4 000 |
| IC 10 | Irr | 11.7 | 4 000 000 | 6 000 |
| GR8 | Irr | 14.6 | 4 000 000 | 200 |
| SagDIG | Irr | 15.6 | 4 000 000 | 5 000 |
| Leo A | Irr | 12.7 | 5 000 000 | 7 000 |
| Dwarf galaxy in Pegasus | Irr | 12.4 | 5 000 000 | 8 000 |
| * Satellites of the Milky Way | | | | |
| # Satellites of M31 | | | | |

The table at left gives information about the Local Group of galaxies. "Type" refers to the Hubble classification, with a few additions: "E" stands for elliptical, "Irr" for irregular, while "dSph" denotes dwarf elliptical galaxies (*dwarf spheroidals*), a recently designated subclass of the ellipticals. Normal spirals are indicated by "S," barred spirals by "SB." The letter "a" indicates spirals with a large nuclear bulge and many narrow, tightly coiled spiral arms; "c" indicates a greatly reduced bulge and wide, very open spiral arms; "b" denotes intermediate types. In 1960, Sidney van den Bergh introduced a further morphological criterion (updated in 1980 by Debra and Bruce Elmegreen). It further subdivides spiral galaxies into "grand design" spirals, in which the arms are long and symmetrical and extend from the nuclear bulge to the outer edge of the disk (for example, M51, shown on page 107); "flocculent" galaxies, in which the arms are made up of short, asymmetrical segments that give a woolly appearance to the structure; and finally, intermediate-type galaxies, with multiple arms, which have two symmetrical arms in the inner portion of the disk, but many branches and arms that are apparently independent in the outer portions (such as M74, shown on page 91).

Most galaxies tend to form groups, or clusters. At right is a diagram showing the group to which the Milky Way belongs, known as the Local Group. Irregular galaxies are shown in blue, spirals in yellow and ellipticals in red.

Our Local Group is fairly modest, comprising only about 30 members, many of which are satellites of the two main galaxies, our own and M31. Its total radius is about 5 million light-years. As with a determination of the structure of our galaxy, it is also difficult to determine exactly which nearby galaxies belong to the Local Group. For example, some members could be hidden behind the plane of the Milky Way, while others could be too diffuse and dim to be glimpsed; as well, the limits of the group are not clearly defined.

*The Local Group is a kind of universe in miniature. Its significance lies in the fact that the galaxies which comprise it are all fairly close to us, close enough to be resolved into structural components—stars, nebulas and clusters—that we can see well. This makes it an excellent testing ground on which to verify our theories about extragalactic distances, galactic evolution, galactic star populations, and so on.*

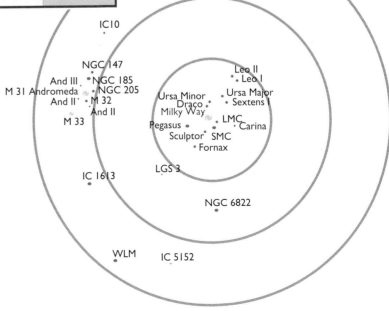

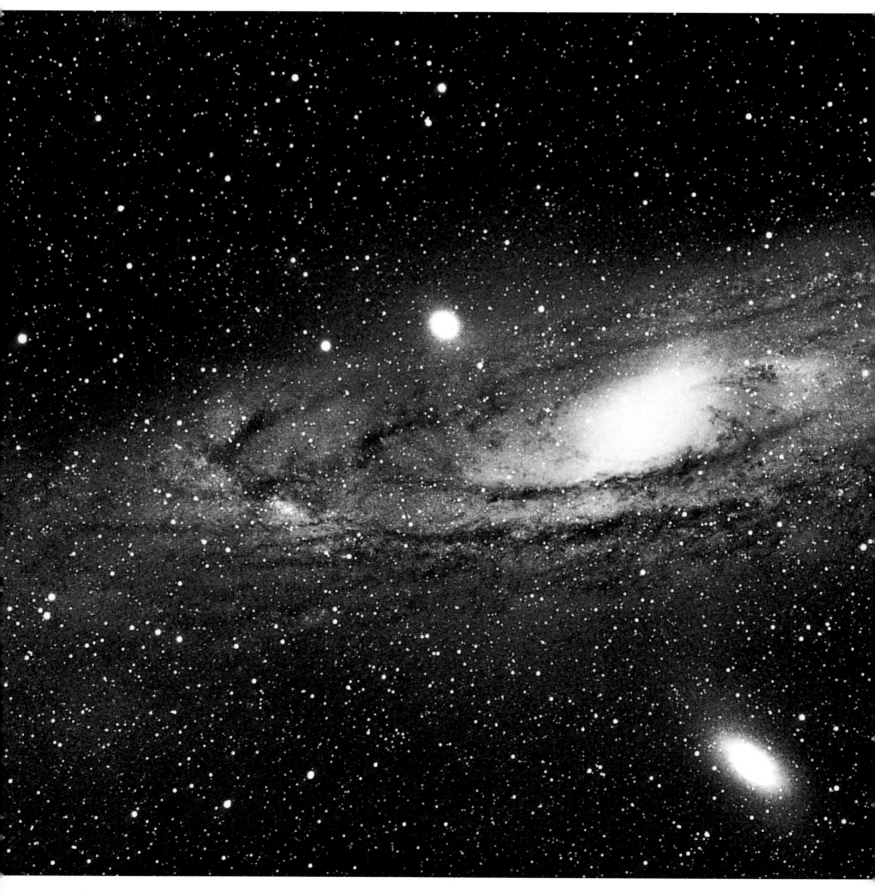

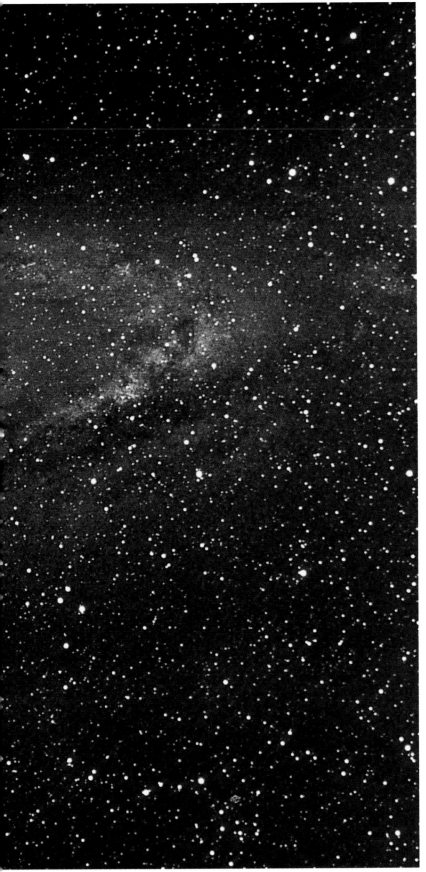

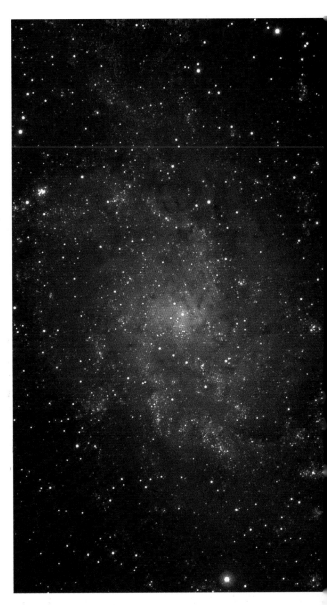

Left, M31, the Andromeda Galaxy, is the largest member of the Local Group. It is a spiral galaxy containing 300 billion stars. Its nuclear bulge is about 12,000 light-years wide. In it lies the true galactic nucleus, which is 35 light-years across, and is double, as the Hubble Space Telescope revealed in 1991. Hidden deep within it, there is probably a black hole about 50 million solar masses in size. It is only in photographs that the spiral structure of the galaxy is visible. Within the structure, Baade has identified no fewer than seven spiral arms. M31 has eight satellite galaxies, two of which—M32 and NGC 205—are visible in this picture.

■ Photograph by Tony Hallas from two Fujicolor Super HG 400 hypersensitized negatives; 50-minute exposures taken with a 15-cm f7.5 Astro-Physics refractor and a Pentax 6 x 7 camera. Two 10 cm x 13 cm transparencies were obtained from the negatives and, from these, two inter-negatives, which, put together, produced the final product.

Above, M33, the Triangulum Galaxy, is the third-largest galaxy in the Local Group, after M31 and our own. Besides the clearly visible spiral structure, note the areas of ionized hydrogen (the red spots) and the peculiar lack of a nuclear bulge.

■ Photographed by W. Schoening and N. Sharp on Ektachrome film at the prime focus of the 3.8-m Mayall Telescope of the KPNO.

*The Andromeda Galaxy was the first galaxy whose distance was measured. In 1924, Edwin Hubble, from a careful study of the Cepheids contained within it, determined its size, from which it was obvious that M31 was a star system that was completely outside and separate from the Milky Way. This was the first step toward the concept of a universe not limited to our own galaxy and much vaster than previously believed.*

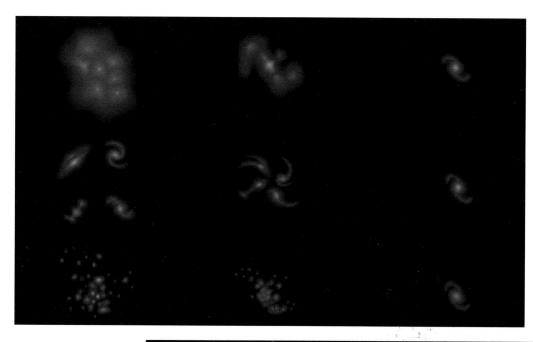

**Left, various models of the formation of spiral galaxies. According to Sandage, Eggen and Lynden-Bell, a spiral galaxy is formed from the rapid collapse of a large rotating cloud of gas (top row). Alar Toomre's model sees spirals as arising from the fusion of numerous large gas structures (middle row). A variation attributed to Searle and Zinn envisions an aggregate starting from smaller, more numerous gas clouds (bottom row). It is probable that the valid theory is a combination of these three: the first could account for the innermost components of the galaxy (the nuclear bulge and inner halo), while the outer halo could have been formed from the fusion of various fragments.**

**Right, M83, one of the most spectacular spiral galaxies, with well-defined arms. In this picture, despite the galaxy's considerable distance from us (14 million light-years), a few particularly bright individual stars, bluish in color, can be seen. The spiral arms also contain many clusters of young, blue stars. The pink areas are gaseous nebulas quite similar to those in our own galaxy.**
**■ Tricolor composite by David Malin, from three plates exposed by him for 30 minutes each, using the AAT.**

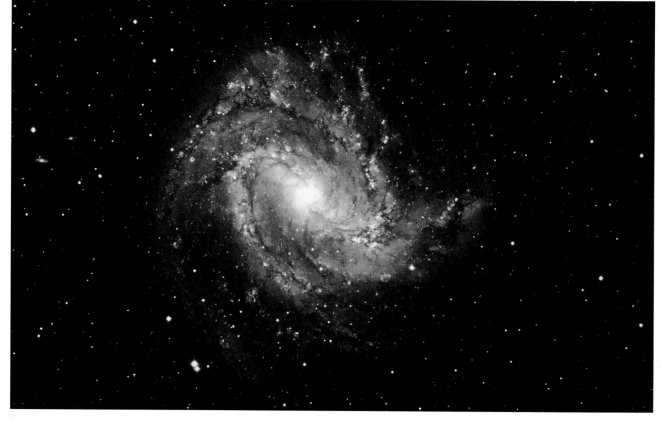

*There are many different theories concerning the origin of the spiral arms of a galaxy. Frank Shu and Chia-Chiao Lin suggested in 1964 that the spiral arms form because of "density waves," grooves or depressions in the galactic disk that are gravitational in origin. In the inner region of the disk, stars move faster than the waves and overtake them; in the outer regions, the opposite occurs. In this way, dust and gases remain trapped in the grooves and are then powerfully compressed by the arrival of other gas carried by the galaxy's rotation. In the end, the pressure generates the spiral arms. Another theory, suggested by Alar and Juri Toomre in 1972, is that in at least a few cases, the formation of the characteristic spiral structure is a result of tidal interaction between galaxies (see page 106). Finally, a theory put forward by Philip Seiden and Humberto Gerola in 1978 suggests that at least the outer spiral structure could be produced by gas instabilities and supernova explosions. These initiate a sequence of star formation and induce warps in the interstellar medium that subsequently degenerate into spiral arcs.*

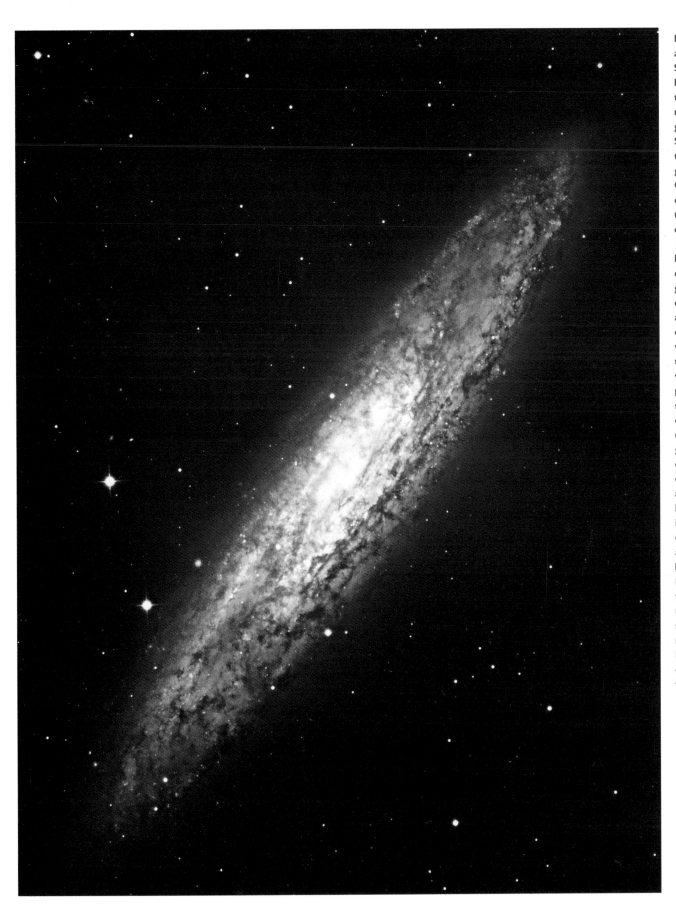

NGC 253, a spiral galaxy in the constellation Sculptor, is 10 million light-years away and is the most conspicuous member of the group of galaxies known as the Sculptor Group. This is the nearest group of galaxies to our Local Group, and like the Local Group, it is part of the Local Supercluster of galaxies (see page 101). NGC 253 is seen here almost edge-on to our line of view. This, together with the presence of the dark lane along its lower right edge (made up of dust which blocks the light of the stars behind it), prevents us from fully appreciating its spiral structure. The presence of the dust lane shows us that this part of the galaxy is the part closest to us. NGC 253 is one of the galaxies that are richest in interstellar dust, and therefore, it is among the most opaque. Its diameter is about 75,000 light-years.

■ Image created by David Malin, using the technique of unsharp masking, from three shots in blue, green and red light taken by Ken Freeman with the AAT; each plate was exposed for 25 minutes.

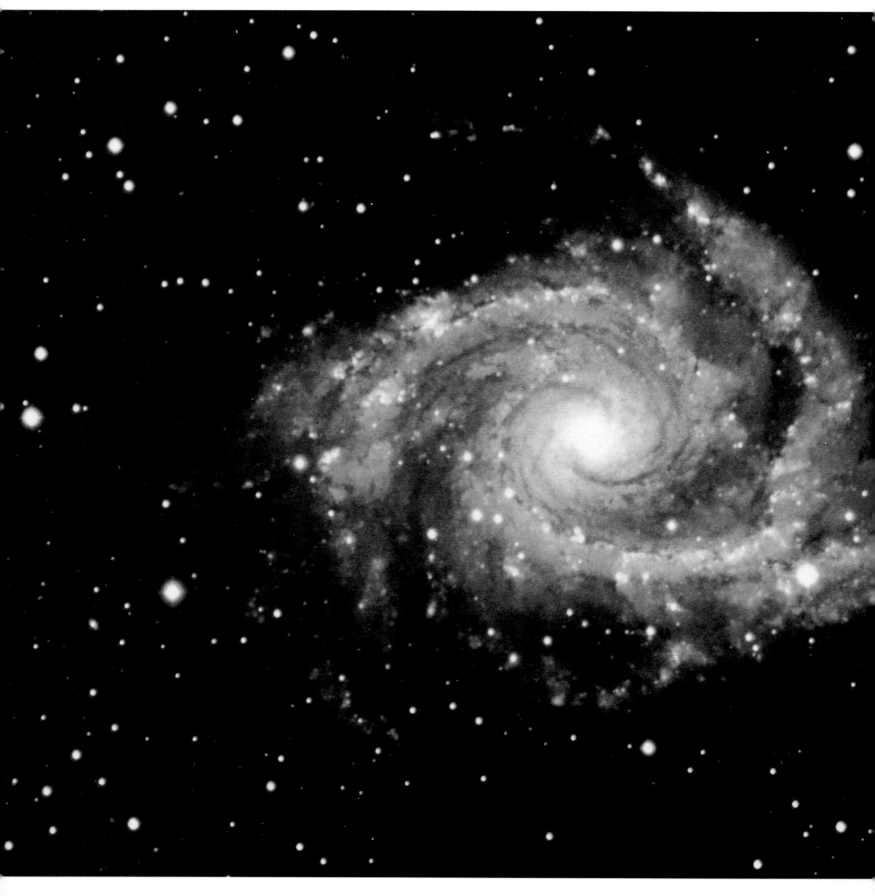

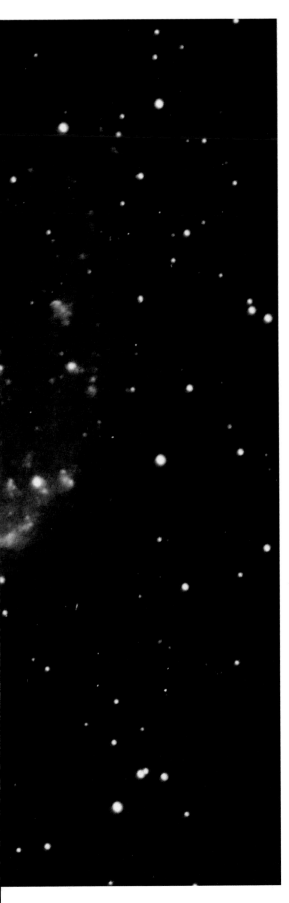

**Left, NGC 2997,** in the constellation Antlia, lying at a distance of 35 million light-years, is one of the most beautiful spiral galaxies in the sky. This photograph gives an excellent indication of the distribution of stars within a typical spiral galaxy. In the arms, it is easy to see the blue coloration of the newer clusters and stars, together with the little pink spots of the nebulas that generated them. In the nucleus, it is also possible to see the gaudy yellow color of the older stars.
■ Tricolor composite by David Malin, from three plates taken by David Carter with the AAT in blue, green and red light, exposed for 30, 25 and 40 minutes, respectively.

**Below, M74,** a type Sc spiral in Pisces, is 30 million light-years away and has a diameter of 80,000 light-years.
This false-color-enhanced image shows the structure and fine details to better advantage.
■ CCD image, using a cold camera; a 30-second exposure taken with the 3.8-m Mayall Telescope of the KPNO.

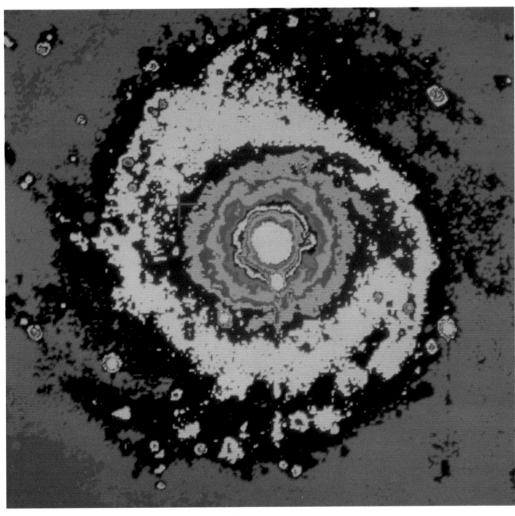

*The first to observe the spiral arms in a distant galaxy was William Parsons, the third Earl of Rosse. A wealthy Irish amateur astronomer, he ordered a gigantic telescope to be built between 1842 and 1844, with a 1.8-m mirror. The telescope, the largest in the world at the time, consisted of a tube 17 m long placed between two walled towers that supported the guides within which it moved. In April 1845, "the Leviathan of Parsonstown,"*

*as it was known, was pointed at the 51st object in Messier's catalog. Until then, M51 had been seen as a round, milky patch with a wide halo around it. Parsons could now clearly observe the structure of two arms that wound outward along a spiral trajectory from what proved to be the center of the "nebula."*

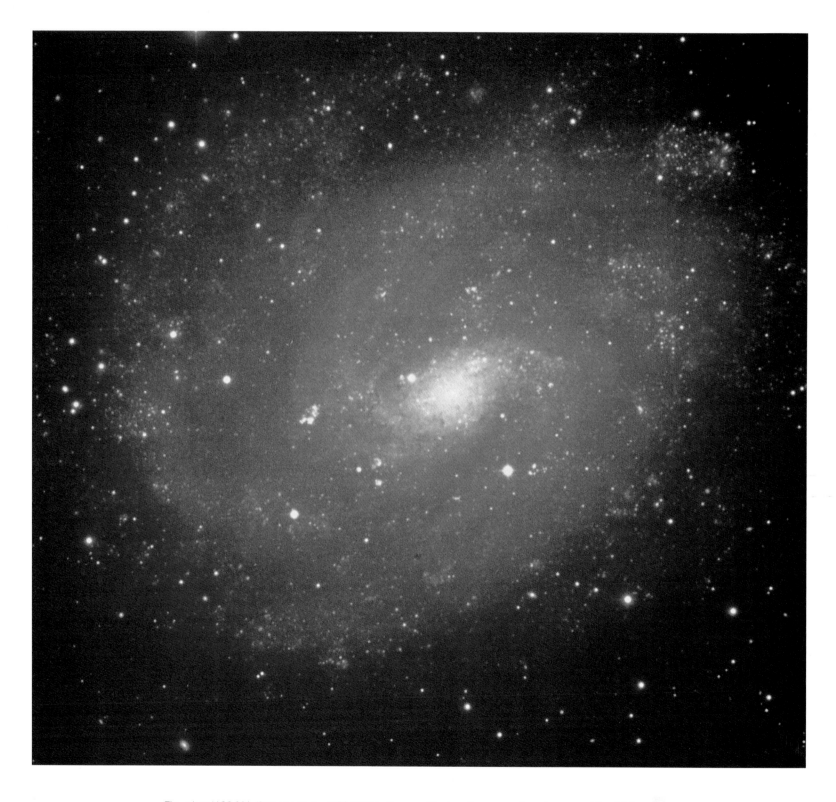

The galaxy NGC 891 (facing page), like NGC 4565 in Coma Berenices, is one of the best examples of a spiral galaxy seen edge-on. Although its fine structure is visible only in long-exposure photographs, the dark dust lane was noticed around the year 1850 by the Earl of Rosse with his "Leviathan" and is clearly visible in a sketch he made at that time. To his contemporary John Herschel, visual observation of the galaxy suggested a "flat, thin ring, of enormous dimensions, seen obliquely." One of the first photographs of NGC 891 was taken by Isaac Roberts, the famous photographer of the Andromeda Galaxy mentioned on page 11. In 1891, he exposed a plate, with his 51-cm reflector, on which the dark lane is visible as clearly as in any modern photograph.

Facing page, NGC 300, a spiral galaxy in the Sculptor Group. A little closer than NGC 253, it is 7 million light-years away. It is similar to M33, and its nucleus is very small. The diffuse nebulas here appear arranged at random rather than aligned along the spiral arms as in other galaxies. NGC 300 is also more transparent than other galaxies, having a low dust content.
■ Tricolor composite image produced from three plates exposed with the ESO's 3.6-m telescope.

Top right, this is how our galaxy would look if it were viewed from outside, along the plane of the galactic disk. NGC 891, a spiral in Andromeda seen edge-on, exhibits a dark band of dust. Although, as in all galaxies, this dust constitutes less than 1 percent of the total galactic mass, it is concentrated mainly within the disk, where it absorbs the light of the stars and nebulas behind.
■ Tricolor composite image by David Malin from plates exposed with the 2.5-m Isaac Newton Telescope, in the Canary Islands, by the INT team, using exposures of 35, 40 and 30 minutes.

Bottom right, the galaxy NGC 7331 in Pegasus is 45 million light-years away and a little larger than the Milky Way, having a diameter of 140,000 light-years.
■ Photographed by William Miller with the 122-cm Schmidt telescope on Mount Palomar.

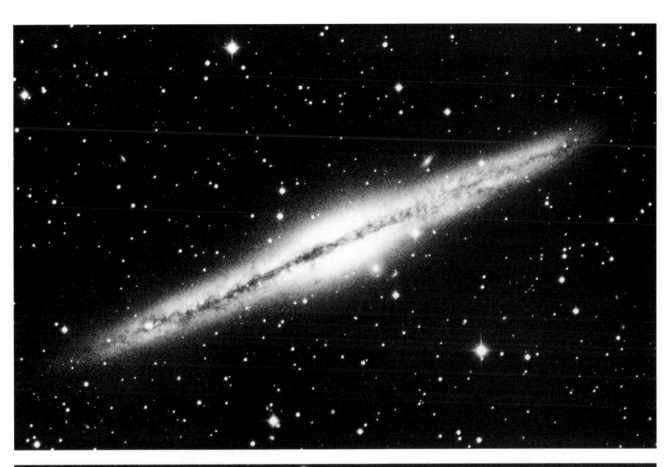

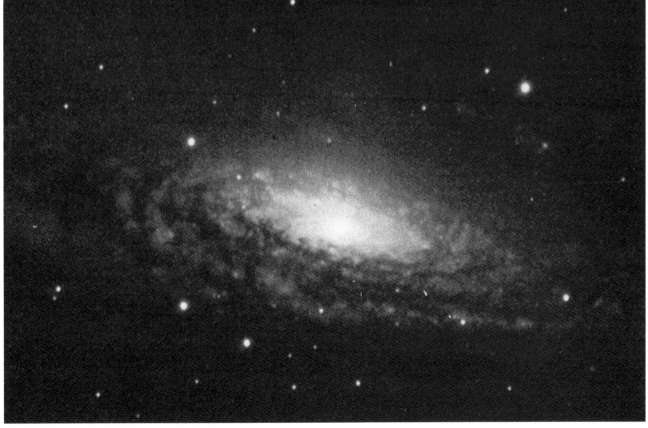

At right, the galaxy NGC 7479 in Pegasus is a fine example of a barred spiral, in which the spiral arms are attached to the ends of a central bar rather than emerging from the nucleus. This galaxy is classified as type SBb+ (with size, shape and structure of its bar and arms intermediate between types b and c). It is approximately 115 million light-years distant, and its diameter is about 135,000 light-years.

■ The image is a composite of several exposures (totaling 55 minutes) taken in near infrared by Boyle, Burg, deJong, Ponder and Windhorst of the University of Arizona with the 1.8-m Vatican-built Advanced Technology Telescope on Mt. Graham, Arizona. It is part of a survey of 500 galaxies carried out in various wavelengths to refine the morphological parameters for classifying types of galaxies.

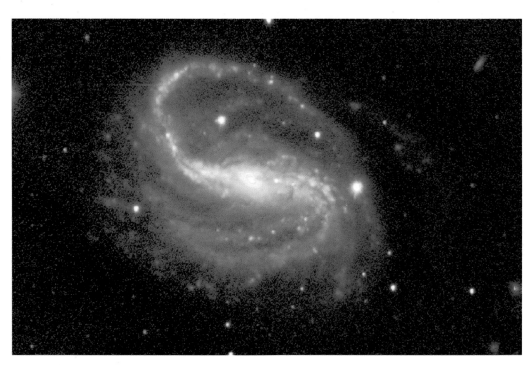

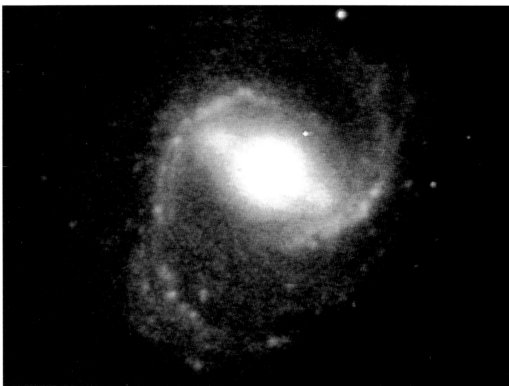

Left, M91, a barred spiral galaxy in the constellation Coma Berenices. For a long time, M91 was a phantom object in Messier's catalog, in the sense that it was believed that the object, observed by the French astronomer in 1781, was actually a comet that had not been recognized as such. It was, however, subsequently identified with the galaxy NGC 4548.

■ Picture taken at the KPNO.

On facing page, NGC 1365, a barred spiral, is the largest galaxy in the Fornax I cluster (see page 102). It measures 170,000 light-years in diameter and is the galaxy that presents probably the best-defined structure of the central bar.

■ Composite image produced from three plates exposed in blue, green and red light with the ESO's 1-m Schmidt telescope.

*The first identification of the barred form of spiral galaxy dates from 1918. In that year, Heber Curtis noticed objects that had "a band of matter stretching diametrically across the nucleus and the inner portions of the spiral." He called them type Ø spirals, because they recalled the shape of the Greek letter theta. It was Hubble, in 1926, who coined the term "barred spirals." He also saw that although these systems seemed to have the same general properties as normal spirals, they were much less numerous. Today, it is believed that about one-third of all spirals are barred.*

*It is not clear precisely how a bar, rather than a central bulge, forms in a disk-shaped galaxy. Probably, the birth of a bar represents an ordered tendency of the stars in a galaxy to arrange themselves within relatively stable orbits, despite their apparently chaotic motion. In this context, the central bar could be the sign of a balance that has been reached, exactly like a tightrope walker holding a balancing pole. The formation of central bars fits in well with Shu and Lin's density-wave theory of the origin of spiral arms.*

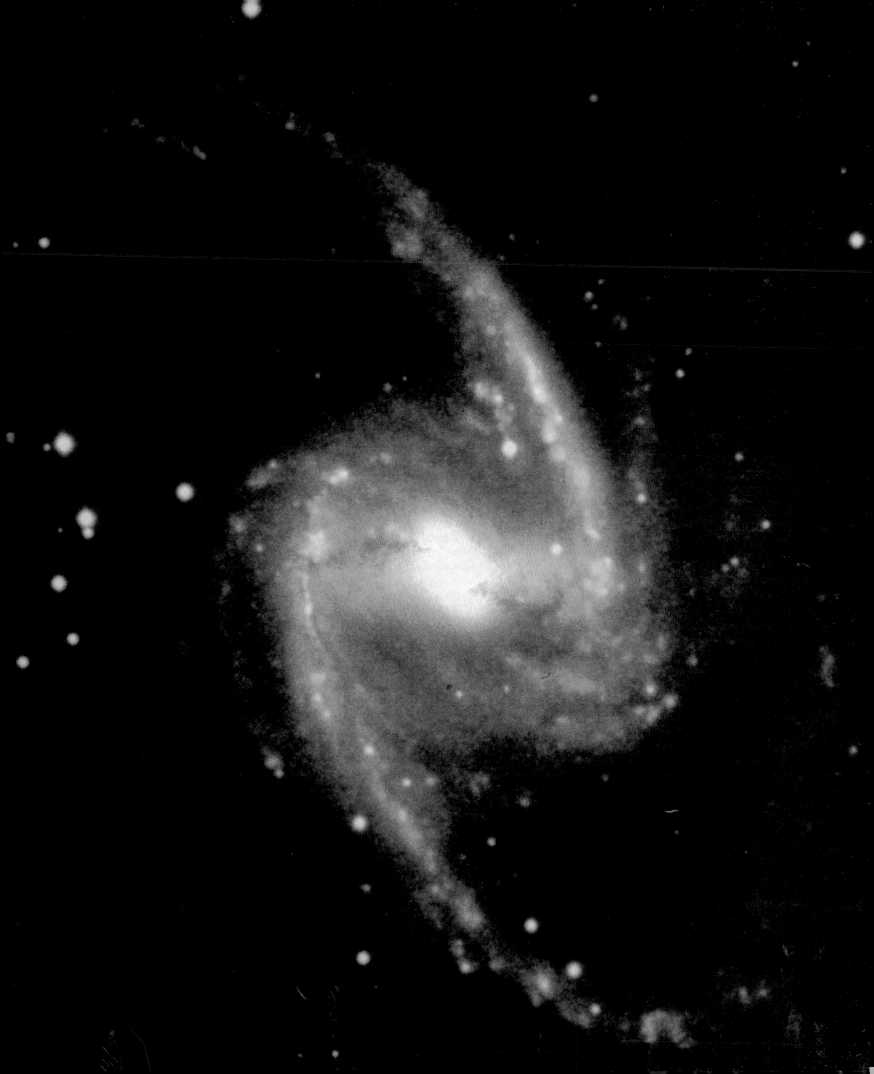

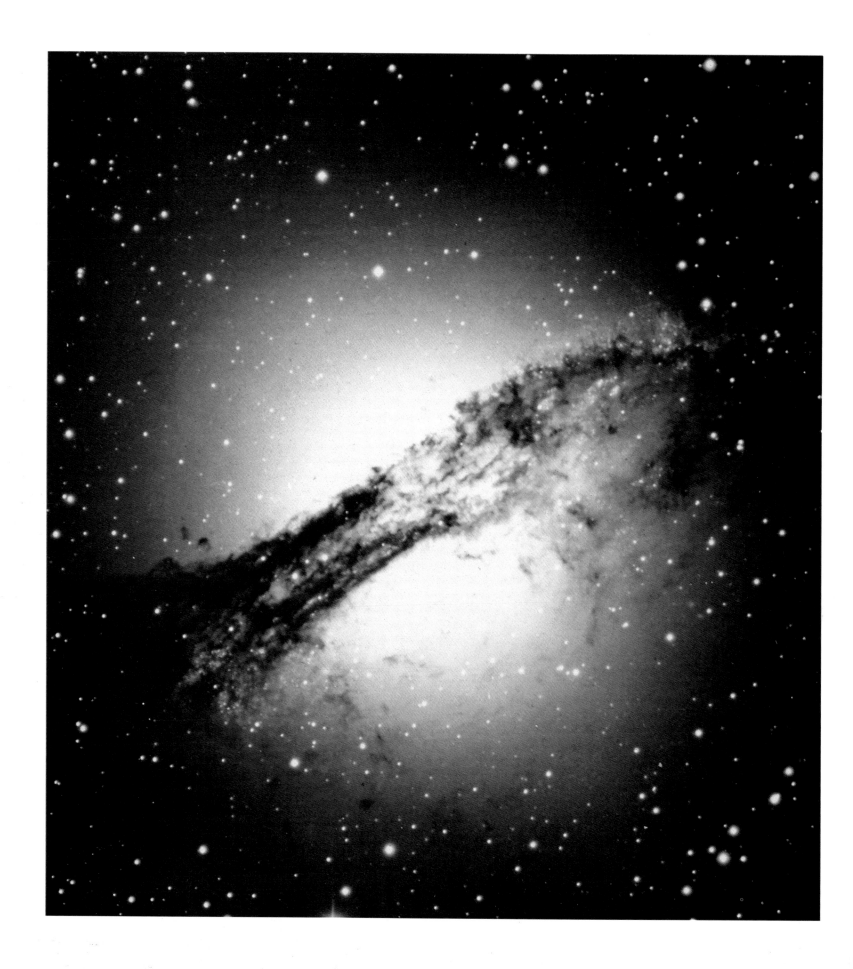

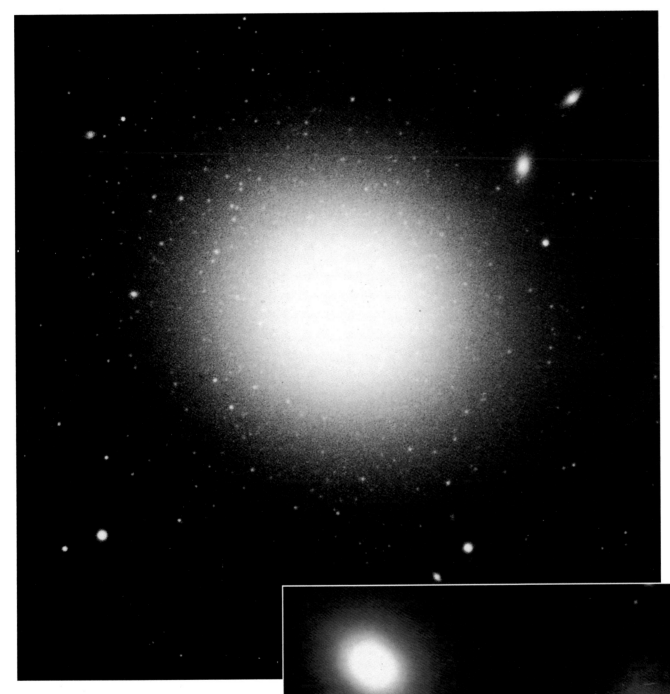

According to recent theories, elliptical galaxies are formed by the fusion of disk-shaped galaxies. Theirs is a violent origin, involving a rapid collapse and a paroxysmic process of star formation, leading to the rapid consumption of all the gas present. This explains why the stars of elliptical galaxies are much older than those of spiral ones, as evidenced by the yellow color of the galaxy M87, visible at left.
■ Photograph by D. Malin, from plates exposed by him and by R. Cannon for 30, 25 and 50 minutes on the AAT.

Below, the elliptical galaxy NGC 1199 (on the left), together with a barred spiral and a lenticular companion. The different colors of these three objects clearly demonstrate the range of diversity in star composition among galaxies.
■ Tricolor composite of three CCD images taken by N. Sharp with the KPNO's 90-cm telescope.

On facing page, NGC 5128 in Centaurus, an elliptical galaxy which is the result of the interaction between two separate galaxies. That this interaction is still occurring is demonstrated by the nebulosity present inside the wide, dark dust band, clearly the center of intense star-forming activity, and by the powerful radio-wave emissions coming from the core of the galaxy (also known as radio source Centaurus A).
■ Tricolor composite image created by David Malin from three plates exposed by him and by Ray Sharples for 85, 90 and 80 minutes on the AAT.

Irregular galaxies are generally rather small and are often found close to large elliptical or spiral galaxies, by which they are gravitationally perturbed. The origin of their amorphous, or chaotic, aspect is said to lie in these perturbations. There are also some isolated and some rather massive irregular galaxies. In irregular galaxies, star formation proceeds at a slower and more constant pace in comparison with the elliptical ones, and they still retain a percentage of the original gas. On this page is a picture of the closest one to us, the prototype of irregular galaxies—the Large Magellanic Cloud, a satellite of the Milky Way. Its proximity and the diversity of its stellar populations in comparison with those of our own galaxy make it an astrophysical laboratory of inestimable value for testing our theories about stellar evolution.
■ Tricolor composite created by David Malin using **UKS** plates exposed for 40, 40 and 60 minutes by **Liz Sim**.

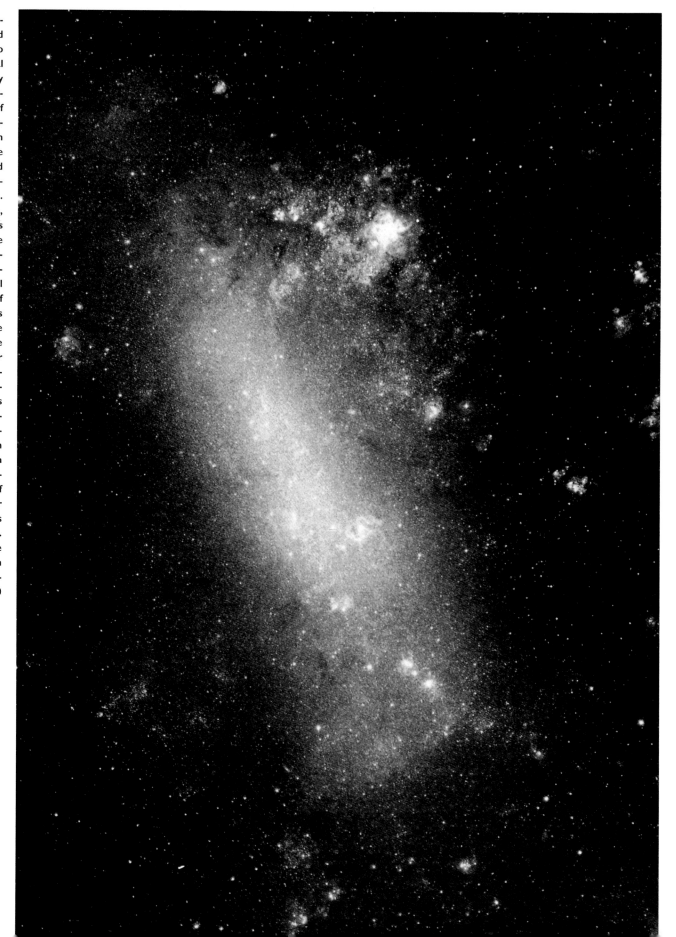

Below, **NGC 55**, in the nearby **Sculptor Group**, has been categorized as both a spiral galaxy viewed edge-on and an irregular galaxy. The latter is, however, the more probable. This galaxy might be similar to the **Large Magellanic Cloud**, but it is difficult to determine its shape because of the unfavorable perspective. **NGC 55** has a kind of bulge or large bar and a diffuse disk. The bulge, unlike that in spiral galaxies, does not occupy a central position but is located at one end of the disk. It would also ap-pear that the galaxy's center of gravity does not coincide with the bulge. The disk has a complex structure, in which a few bright stars can easily be resolved. Its diameter is about 40,000 light-years.

■ Using the ESO's 3.6-m telescope, Richard West produced this tri-color composite image from three plates exposed in blue, green and red light (for 45, 45 and 60 minutes).

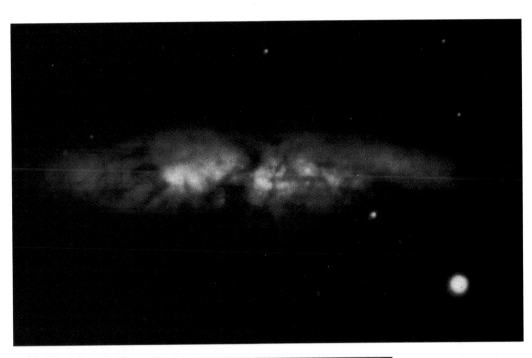

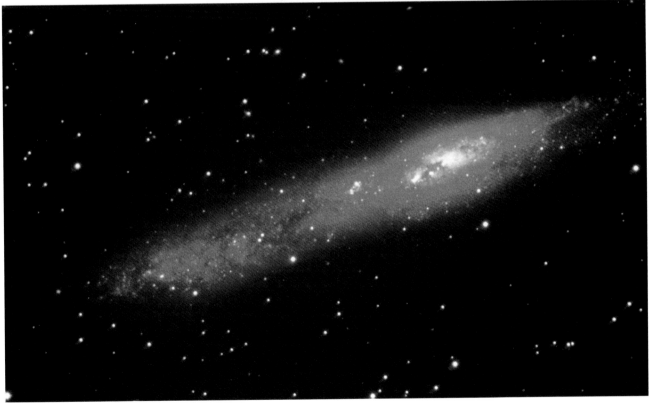

Above, **M82 in Ursa Major**, a prototype of the Class II irregular galaxies, characterized by a highly chaotic structure and by the presence of bright filaments that seem to emerge from the nucleus. **M82**, which is also a radio source, is strongly perturbed by the nearby galaxy **M81**. M82 lies at a distance of 16 million light-years and has a diameter of 50,000 light-years.

■ Photographed by **William Miller** with the 508-cm Hale Telescope on Mount Palomar.

*The Magellanic Clouds are the two most precious pearls in the southern sky. Even a traveler uninitiated in astronomical matters can glimpse them with the unaided eye as two diffuse patches, almost like two pieces of the Milky Way. They were undoubtedly noticed by the Portuguese on their way to the Cape of Good Hope and by Vespucci, the explorer of South America. The first to write about them was Florentine sailor Andrea Corsali in 1517:*

*"We saw two clouds of considerable splendor moving around the Pole, rising and ebbing continually, maintaining a circular motion." Their name is linked especially to Antonio Pigafetta's Relazione ("Report"). During Magellan's voyage, the Italian sailor wrote in December 1520: "Many small stars are visible close together, almost like two clouds, not much separated from one another and slightly obscured."*

A region near the center of the cluster of galaxies known as the Virgo cluster, which, being the closest large cluster to us, occupies quite a wide area of the sky. This is an irregular cluster, with members that are widely separated and with no central concentration, which comprises both spiral galaxies (about 70 percent) and elliptical ones. The area where the galaxies are most concentrated is shown here, including the giant elliptical galaxies M84 (on the right) and M86 (center). The dynamic center of the cluster is, however, located near the most massive galaxy, M87 (see page 97).

Visible in the picture are edge-on spiral galaxies, dwarf elliptical galaxies and interacting galaxies.

■ Tricolor composite image created by David Malin from three plates (exposed in red, green and blue light for 30 minutes each) taken by John Barrow with the UKS.

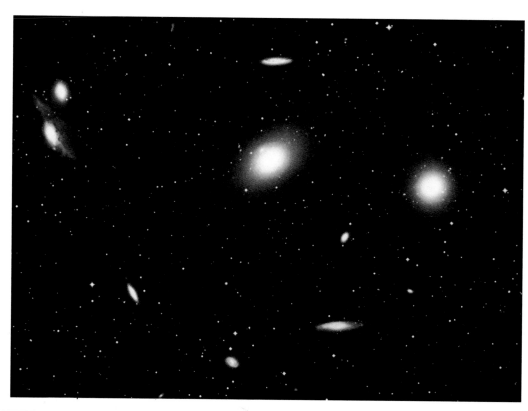

| NAME | DISTANCE (LIGHT-YEARS) | NUMBER OF GALAXIES | DIAMETER (LIGHT-YEARS) |
|---|---|---|---|
| Virgo | 55 000 000 | 2 500 | 14 000 000 |
| Fornax I | 60 000 000 | | 7 000 000 |
| Fornax II | 65 000 000 | | 8 000 000 |
| Puppis | 70 000 000 | | |
| Hydra I | 120 000 000 | | |
| Centaurus | 130 000 000 | 300 | 7 000 000 |
| Pegasus I | 160 000 000 | 100 | 3 000 000 |
| Cancer | 200 000 000 | 150 | 15 000 000 |
| Perseus | 220 000 000 | 500 | 15 000 000 |
| Abell 194 | 220 000 000 | | 1 000 000 |
| Abell 1367 | 250 000 000 | | |
| Coma Berenices | 270 000 000 | 800 | 25 000 000 |
| Abell 400 | 290 000 000 | | |
| Abell 2197 | 370 000 000 | | |
| Abell 2199 | 375 000 000 | | 1 000 000 |
| Leo A | 430 000 000 | | |
| Hercules | 450 000 000 | 300 | 1 000 000 |
| Abell 2152 | 470 000 000 | | |
| Pegasus II | 520 000 000 | | 25 000 000 |
| Ursa Major I | 620 000 000 | 300 | 10 000 000 |
| Haufen A | 650 000 000 | | |
| Leo | 800 000 000 | 300 | 10 000 000 |
| Corona Borealis | 880 000 000 | 400 | 10 000 000 |
| Gemini | 950 000 000 | 200 | 10 000 000 |
| Boötes | 1 600 000 000 | 150 | 12 000 000 |
| Ursa Major II | 1 700 000 000 | 200 | 12 000 000 |
| Hydra II | 2 500 000 000 | | |

A table of information about the main galaxy clusters known, listed in order of increasing distance from us. Galaxy clusters are of different types, rather like the galaxies themselves. There are spherical clusters (similar in structure to an elliptical galaxy), in which the most important members are elliptical and S0 galaxies. A typical example is the Coma cluster. Then there are irregular clusters, such as the Virgo cluster (above), in which the distribution of the galaxies (mainly spirals) is chaotic, like that of the stars in the giant irregular galaxies. Finally, there are intermediate clusters (similar in many ways to the structure of disk-shaped galaxies), which often consist of groups of elliptical galaxies in the center plus an extended peripheral halo of other galaxies. These are made up of galaxies of every type, without any particular distinction. The size of spherical and intermediate clusters is on the order of a few tens of millions of light-years, while irregular ones are highly variable in size and can even be quite small.

*Even before galaxies were identified as systems that were independent from, and completely external to, the Milky Way, their tendency to form groups within certain parts of the sky was noticed. As far back as 1933, Harlow Shapley published a first catalog of 25 galaxy clusters. In 1958, George Abell published a list of 2,712 clusters, and between 1960 and 1968, Fritz Zwicky published a work in six volumes that listed 9,134 clusters.*

On facing page, a diagram illustrating the more than 50 groups of galaxies, analogous to our own Local Group, cataloged by French-American astronomer Gérard de Vaucouleurs, within a radius of 50 million light-years. The groups are shown as spheres of a diameter proportional to their dimensions. Each group is indicated by a progressive number and by either the name of the constellation in which it appears or the name of its main galaxy. The group at the center is the Local Group. Note the concentration of groups toward the right. These are part of a larger grouping, known as the Local Supercluster, which is centered around the Virgo cluster of galaxies.

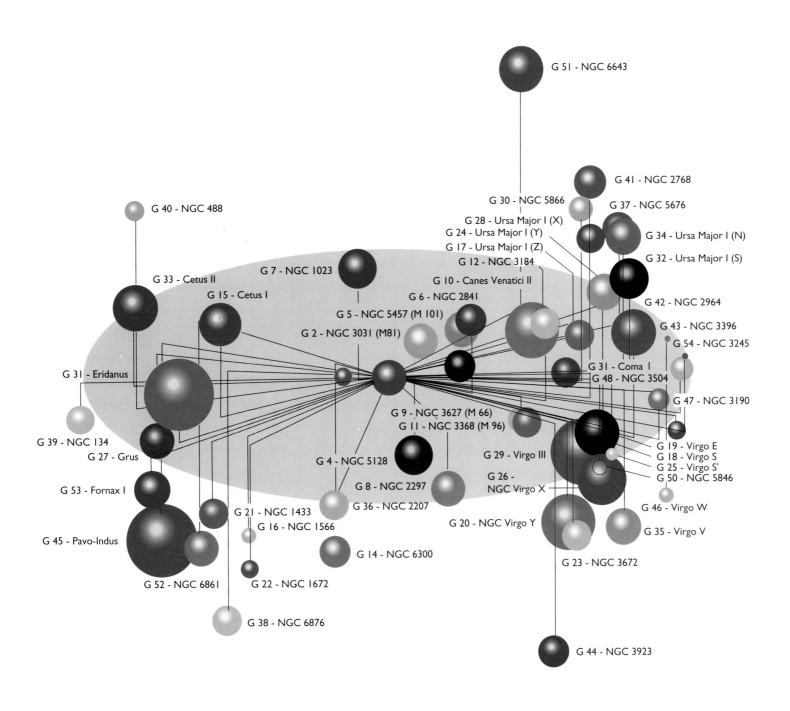

G 51 - NGC 6643

G 41 - NGC 2768

G 30 - NGC 5866

G 37 - NGC 5676

G 28 - Ursa Major I (X)

G 24 - Ursa Major I (Y)

G 34 - Ursa Major I (N)

G 17 - Ursa Major I (Z)

G 32 - Ursa Major I (S)

G 12 - NGC 3184

G 40 - NGC 488

G 33 - Cetus II

G 7 - NGC 1023

G 10 - Canes Venatici II

G 42 - NGC 2964

G 15 - Cetus I

G 6 - NGC 2841

G 43 - NGC 3396

G 5 - NGC 5457 (M 101)

G 54 - NGC 3245

G 2 - NGC 3031 (M 81)

G 31 - Eridanus

G 31 - Coma I

G 48 - NGC 3504

G 47 - NGC 3190

G 39 - NGC 134

G 9 - NGC 3627 (M 66)

G 11 - NGC 3368 (M 96)

G 19 - Virgo E

G 27 - Grus

G 18 - Virgo S

G 29 - Virgo III

G 25 - Virgo S'

G 50 - NGC 5846

G 53 - Fornax I

G 4 - NGC 5128

G 26 -
NGC Virgo X

G 8 - NGC 2297

G 46 - Virgo W

G 21 - NGC 1433

G 36 - NGC 2207

G 45 - Pavo-Indus

G 16 - NGC 1566

G 20 - NGC Virgo Y

G 35 - Virgo V

G 14 - NGC 6300

G 52 - NGC 6861

G 22 - NGC 1672

G 23 - NGC 3672

G 38 - NGC 6876

G 44 - NGC 3923

Traces of the existence of the Local Supercluster of galaxies can be seen in the early catalogs of nebular bodies produced by William Herschel and his son John during the first half of the 19th century. In these catalogs, a concentration of "nebulas" in the northern galactic hemisphere became clearly evident, due to the peripheral position of our own galaxy in relation to the nearer ones. During the 1920s, Knut Lundmark and J.H. Reynolds noticed a considerable concentration of "nebulas" (galaxies were called "nebulas" in those days), greatest along the celestial equator, but attributed no particular importance to them. Gérard de Vaucouleurs was the first to suggest that such a concentration was the trace of a flattened supersystem, namely the Local Supercluster, centered around the Virgo cluster and containing several dozen galaxy groups as well as individual galaxies. Today, it is believed that the Local Supercluster has a diameter of about 100 million light-years, with a nucleus formed by no fewer than 11 clusters of galaxies and a halo made up of about 50 other clusters plus millions of single galaxies. It has also been theorized that the Local Supercluster is connected to other distant superclusters by means of bridges of gaseous matter.

The Fornax I cluster of galaxies, the second-closest large cluster after the Virgo cluster. This is a spherical cluster, dominated by the presence of elliptical galaxies such as the luminous **NGC 1399** (the brightest) and **NGC 1404** (at top). Also notable is the barred spiral **NGC 1365** (shown in detail on page 95), at lower left in the picture. Both in this picture and in that of the Virgo cluster on page 100, it is evident that the color photography of very distant objects loses value because of the telescope's difficulty in distinguishing individual star populations inside galaxies. ■ David Malin produced this picture by combining two plates taken with the **UKS** by Peter Standen (in blue and green light, with exposures of 40 and 50 minutes) and one by Malcolm Hartley (in red light, with a 60-minute exposure). Using a plate taken in blue light with a 90-minute exposure and applying the unsharp masking technique, Malin has revealed around **NGC 1399** many globular clusters belonging to the galaxy (barely visible in this picture), as well as many background galaxies.

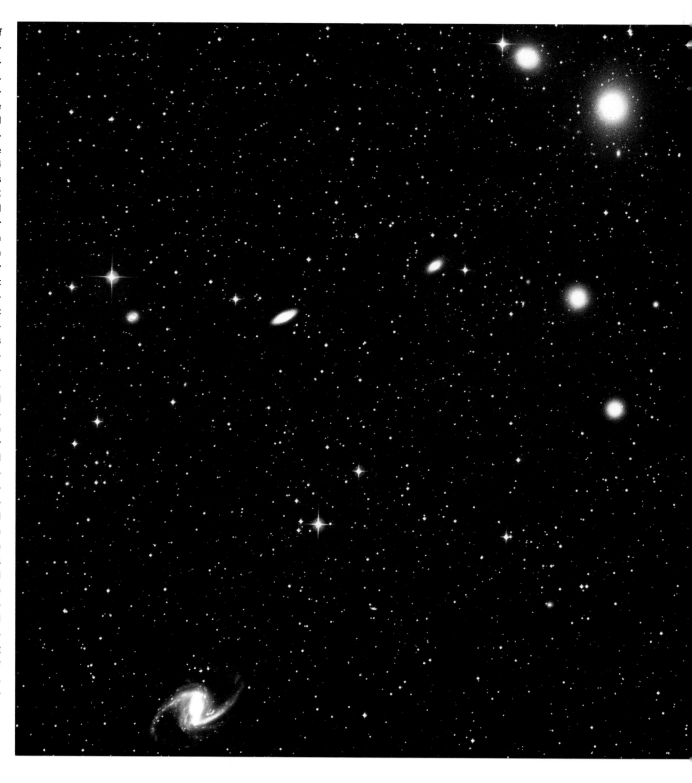

The constellation Fornax was introduced by French astronomer Nicolas Louis de Lacaille in the 18th century, along with 13 other new constellations, the result of the final cartography of the southern sky. These discoveries brought the total number of constellations to the current 88. The work was carried out on Table Mountain, near the Cape of Good Hope. In addition to the creation of a planisphere containing the new constellations, it led to the compilation of a hefty new catalog of 10,000 southern stars. Thanks to this massive work, Lacaille was dubbed the "Columbus of the southern skies." As a tribute to the illuminating spirit of the period, the names of the new constellations were inspired by scientific instruments, technology and the arts. For example, besides Fornax (Furnace), there are Pyxis (Compass), Sculptor, Horologium (Clock), Telescopium, Microscopium and Antlia (Air Pump).

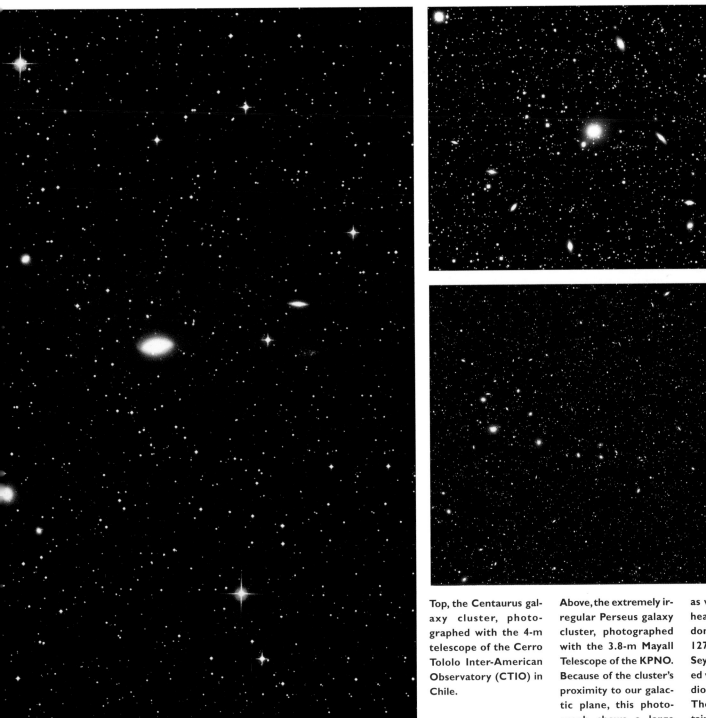

Top, the Centaurus galaxy cluster, photographed with the 4-m telescope of the Cerro Tololo Inter-American Observatory (CTIO) in Chile.

Above, the extremely irregular Perseus galaxy cluster, photographed with the 3.8-m Mayall Telescope of the KPNO. Because of the cluster's proximity to our galactic plane, this photograph shows a large number of star images, as well as galaxies. The heart of the cluster is dominated by NGC 1275 (see page 127), a Seyfert galaxy associated with the powerful radio source Perseus A. The cluster also contains several other radio sources.

The reader should be advised that when galaxies and galaxy clusters are referred to as being "in" a particular constellation, this is not meant to suggest a physical link between the galaxy or cluster and that constellation. Although galaxies and galaxy clusters may appear to lie in the same part of the sky as a given constellation, they are infinitely more distant than the visible stars in the constellation, which are all part of our own galaxy and are no more than a few hundred light-years away. Indeed, the constellations are themselves a mere effect of perspective, because the stars of which they are made up are actually at widely different distances from us. In looking at the sky, we cannot in any way comprehend its "depth"; we see essentially what the ancients saw. But unlike us, they believed that the "fixed stars" were all situated at the same distance from us because they were attached to a solid, crystalline sphere. Certainly, many names of clusters and galaxies have been assigned in periods in which it was not known that such objects were in fact external to our own galaxy. But even in later years, astronomers have found it convenient to assign the names of constellations to distant galaxies and clusters to place them in a context based on the observer's point of view.

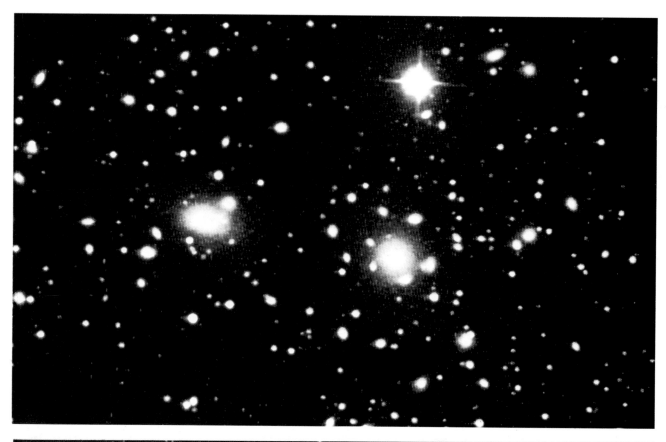

The supergiant elliptical galaxies **NGC 4889** and **NGC 4874**, together with 300 smaller ellipticals, at the center of the Coma cluster. This is a typical spherical cluster with a strong central concentration, that is, with a structure similar to an elliptical galaxy, though on a larger scale. The spirals are in peripheral positions.
■ Image obtained with the KPNO's 3.8-m Mayall Telescope, computer-enhanced to show the true colors of the galaxies.

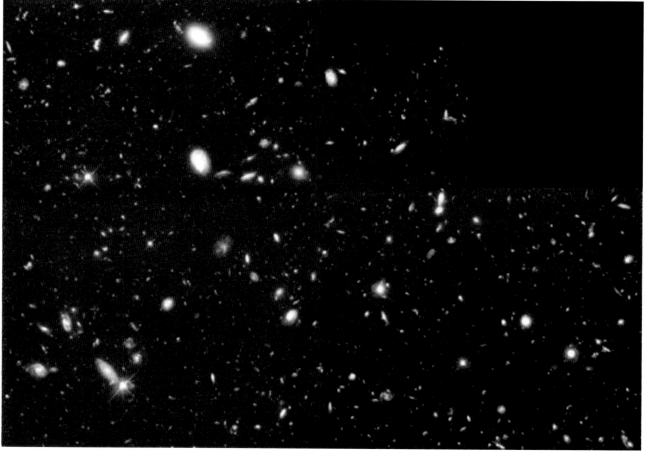

In a region of the sky 1/30 the apparent diameter of the Moon, where ground-based telescopes see only blackness, Hubble captured hundreds of galaxies in this image that portrays the deepest view ever of the universe. Some of the faintest galaxies visible (magnitude 30!) may have formed just one billion years after the Big Bang.
■ The image was assembled from 276 exposures in red, infrared and blue light taken with the **Wide Field/ Planetary Camera 2 (WF/PC-2)** over 10 consecutive days, from December 18 to 28, 1995.

Right, a diagram illustrating the distribution of a few thousand galaxies in the northern hemisphere (top) and the southern hemisphere (bottom). The northern hemisphere map was created by Margaret Geller and John Huchra of the Harvard Smithsonian Center for Astrophysics, while the southern hemisphere map was prepared by Louis Nicolaci da Costa. The work includes about 6,000 galaxies in the northern hemisphere and 3,000 in the south, taking up a volume of space within a 500-million-light-year radius of the Earth. This is, thus, not the entire universe by any means, but nevertheless a significant fraction of it. First, one notices that contrary to what one might have expected, the galaxies are not arranged in a homogeneous way but, rather, are distributed in dense or thin filaments and around vast voids. To the north, for example, a structure known as the "Great Wall" cuts transversely across the area of sky investigated for 500 million light-years. A similar structure, of the same size, can be seen in the southern part; it surrounds a few vast voids and has been called the "Southern Wall."

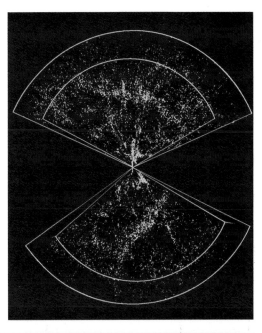

Below, a map drawn in 1990 following 6 months of observations, at a wavelength of 5.7 mm, with the differential microwave radiometer (DMR) onboard NASA's COBE satellite. In a sense, the map represents the oldest fossil of all; that is, the appearance of the universe about half a million years after the Big Bang that gave rise to it, at which time it became sufficiently cold to be transparent to light. The blue and pink areas are relatively cooler and warmer regions of the primordial universe, indicating slight fluctuations that could have been the basis of temperature anisotropies which led later to the formation of the aggregation nuclei from which galaxies and galaxy clusters might have been formed.

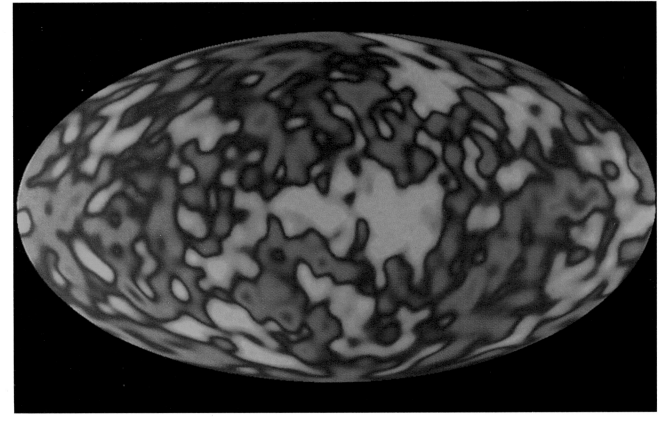

*The COBE map (above) reveals many slight fluctuations. This supports the theory of an inflationary universe, according to which the Big Bang occurred at an extremely rapid pace, so much so as to smooth out the universe, eliminating the larger fluctuations and irregularities. To explain how galaxies and galaxy clusters then formed, it is necessary to hypothesize the intervention, at a given moment, of quantum fluctuations, comparable in general terms to ripples on the surface of a pond. These fluctuations may have constituted the "seeds" for the development of stars and galaxies. An alternative interpretation is that faults in the structure of the expanding space could have been created, either as "cosmic strings" (comparable to the differences in the crystallization of ice between areas of a lake that freeze over at different rates) or as analogous (but more lo-calized) knot-shaped faults known as "intersections." COBE and other instruments have as yet been unable to detect a pattern of fluctuations in the microwave background that would accord fully with either of these theories.*

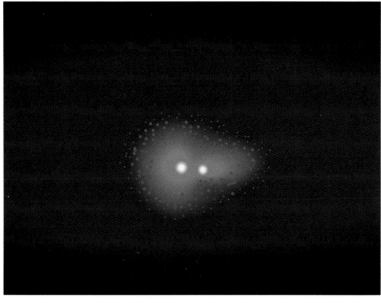 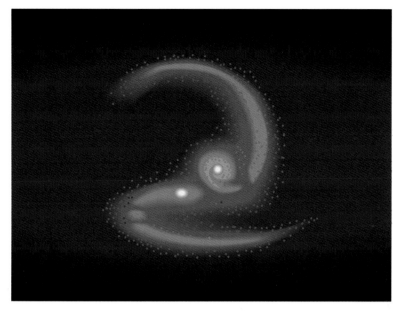

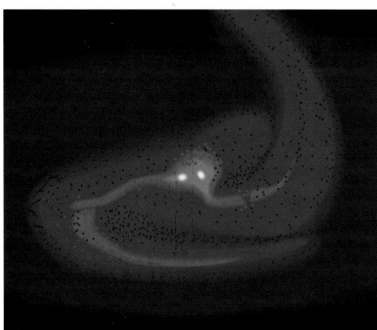 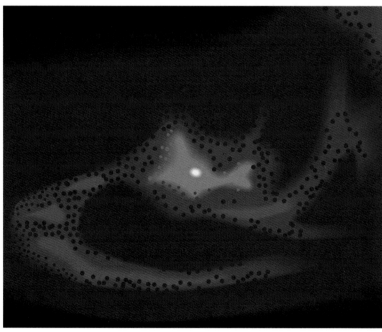

We are now convinced that interaction and fusion between galaxies are events which are more common than was once thought and that they are the basis of the formation of various peculiar galaxies and of various structures, including spiral arms, within the galaxies. Above, a few stages in the fusion of two spiral galaxies. As the two objects gradually draw closer, tidal forces deform the galactic structures, creating filaments and tails. At first, the galaxies draw near in a close embrace, then they orbit for a while, bound together in a kind of balletic pas de deux, and finally, they blend into one.

Facing page, M51, the Whirlpool Galaxy, in the constellation Canes Venatici, and its interacting companion, NGC 5195.

■ Image created by T. Boroson, a composite of three CCD images photographed in red, green and blue light with the 0.9-m telescope of the KPNO.

*A few million years ago, M51 and NGC 5195 underwent a strong interaction that had a powerful influence on both their histories. NGC 5195 is also a spiral, although it does not look like one because it is viewed edge-on and is now located beyond M51, having overtaken it during their close encounter. The most evident tidal distortion wrought on M51 by the passage is the elongation of the spiral arm below, stretching toward NGC 5195. Moreover, it seems that the entire spiral structure could have been produced by the encounter. First, the orbits, initially circular, of M51's stars were made elliptical, and then, because of differential speeds of revolution (the stars closest to the galaxy's nucleus revolve around the core at greater speeds than the outer ones), the stars were grouped into symmetrical spiral structures. The mutual gravitational action of the stars, added to the tidal influence of the companion galaxy, could have considerably amplified such effects, creating a spiral-shaped structure even inside the galactic nucleus. In 1993, Dennis Zaritsky of the Carnegie Observatories and his colleagues found a spiral structure in M51 that can be traced right to the center of the galaxy for about three turns, as well as a central minibar that could have been generated by the tidal interaction with NGC 5195.*

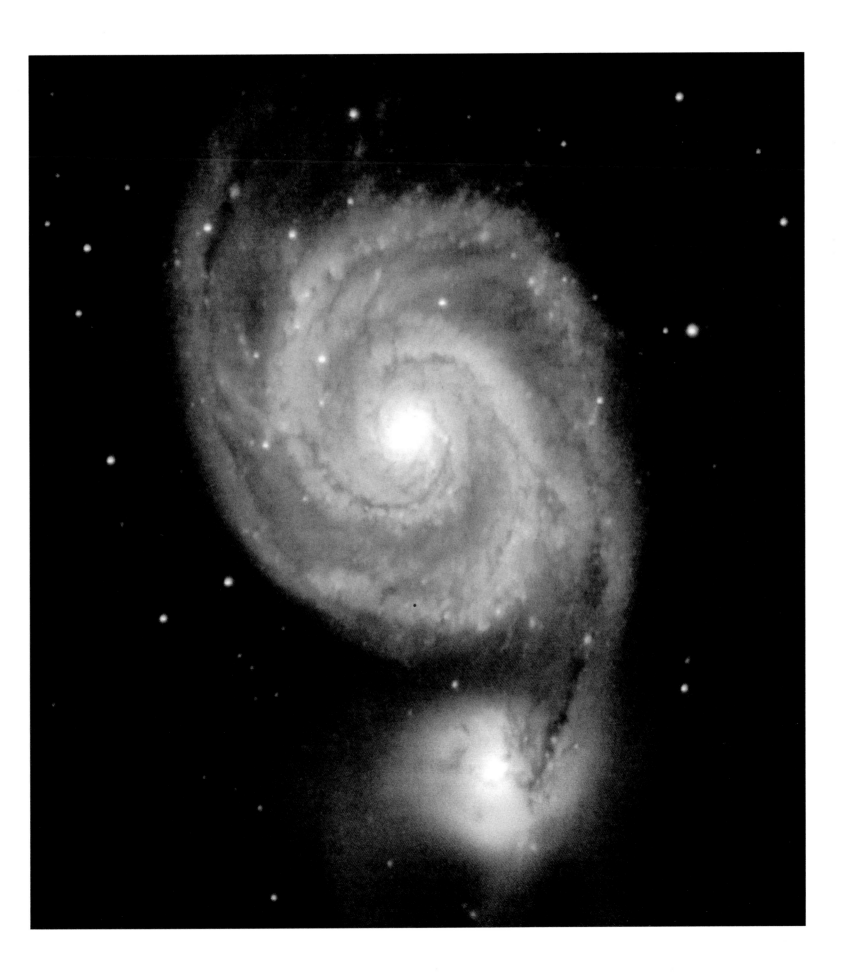

Interacting galaxies **NGC 5394/5395.** The first is a small spiral seen almost face-on, with a very bright nucleus that has an almost starlike appearance, surrounded by surprisingly luminous arcs. The second, on the other hand, is a very tilted spiral. Their distance is 110 million light-years, and their diameters are 50,000 light-years and 100,000 light-years, respectively.

■ Image created by **Nigel Sharp** by combining three **CCD** images photographed in red, green and blue light with the 0.9-m telescope of the KPNO.

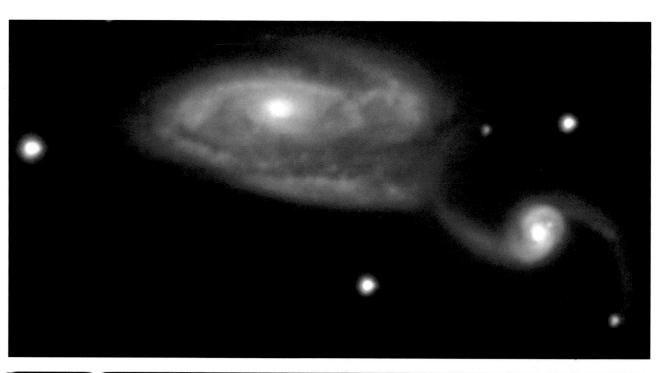

The **NGC 7253** group of interacting galaxies. This is a multiple system that possesses an immense quantity of matter, dust and gas spread among the galaxies. The component at the top, which at first sight appears to be two galaxies, could actually be a chain of three systems in contact with one other.

■ Computer tricolor composite image created by **W. Schoening** and **N. Sharp** from three **CCD** images taken in red, green and blue light with the 3.8-m Mayall Telescope at Kitt Peak National Observatory.

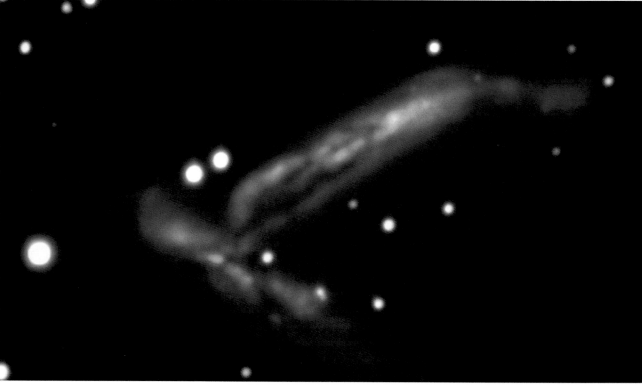

*The existence of galaxies that we could call "peculiar," surrounded by thin filamentary structures such as bridges and tails, has been known since the 1950s, but for a long time, they were considered anomalies. Astronomers felt that phenomena such as collisions between galaxies were highly improbable. There was a decisive change of mind in 1983, when the IRAS satellite revealed hundreds of galaxies with an excess of infrared emissions due to waves of star-forming activity (see page 48), produced by a fusion or collision between galaxies. Interaction between galaxies can also lead to the production of even stranger structures, such as spiral arms, rings of gas and stars and enormous gas shells. The collisions are also at the root of the paroxysmic rate of star formation inside peculiar galaxies known as "starbursts," in which stars are born at the staggering rate of hundreds per year, compared with the two or three per year in a normal galaxy such as the Milky Way. (Star-forming impulses last about 100 million years, after which the galaxy reverts to normal luminosity.) Finally, tidal interactions could be the basis of, or at least favor, the activity of powerfully energetic objects such as quasars and active galactic nuclei.*

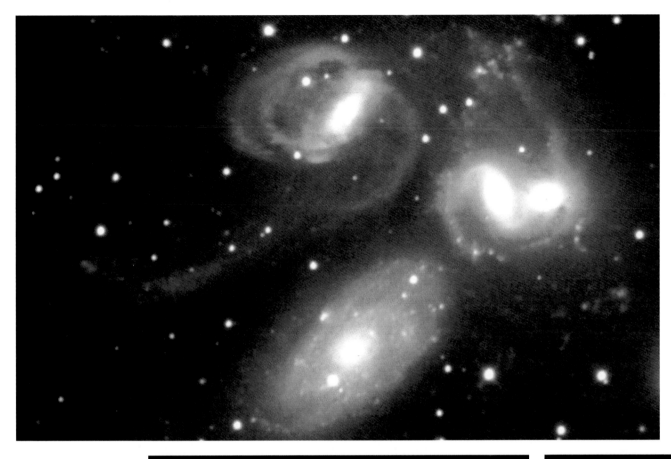

Left, visible here are four of the five galaxies—**NGC 7317, 7318a, 7318b, 7319** and **7320**—which constitute the famous **Stephan's Quintet,** a group of interacting galaxies in the constellation Pegasus (the fifth member lies out of the narrow field of this **CCD** image). One galaxy in the quintet—**NGC 7320**—is not really connected to the group; it is just **40 million** light-years from us, whereas the others are, on average, a good **300 million** light-years away.

■ Picture created by **W. Schoening** and **N. Sharp** by combining three **CCD** images photographed in red, green and blue light with the **3.8-m Mayall Telescope** of the **KPNO**.

Right, the "Cartwheel," a spectacular interacting galaxy, in a picture taken by the Hubble Space Telescope. In ancient times, it was a spiral, its strange current shape having been produced when a small galaxy (out of field) traversed it centrally. The blue areas indicate zones of energetic star-forming activity, while the yellow present in the nucleus and in the inner ring is due to interstellar dust. Between the two rings, there are fragments that could be remnants of the old spiral arms or the embryos of newly forming arms.

■ False-color image created from photographs taken in blue and infrared light with the **WF/PC-2,** by Kirk Borne of the **STScI.**

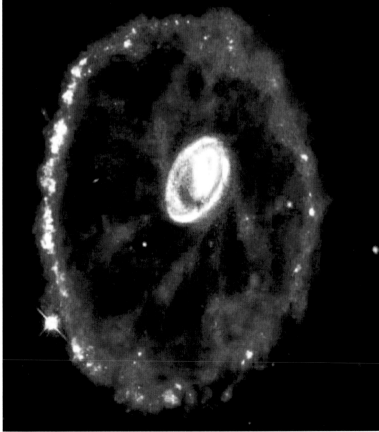

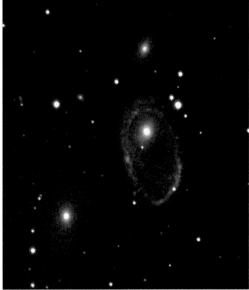

Above, the southern Ring Galaxy, which is located close to the **Large Magellanic Cloud** in the sky. This object is, however, **270 million** light-years away, and it is the product of the emptying of a galaxy with a normal disk by an encounter with a very compact galaxy. Its dimensions are **125,000** by **65,000** light-years.

■ Image created at the **CTIO** in Chile.

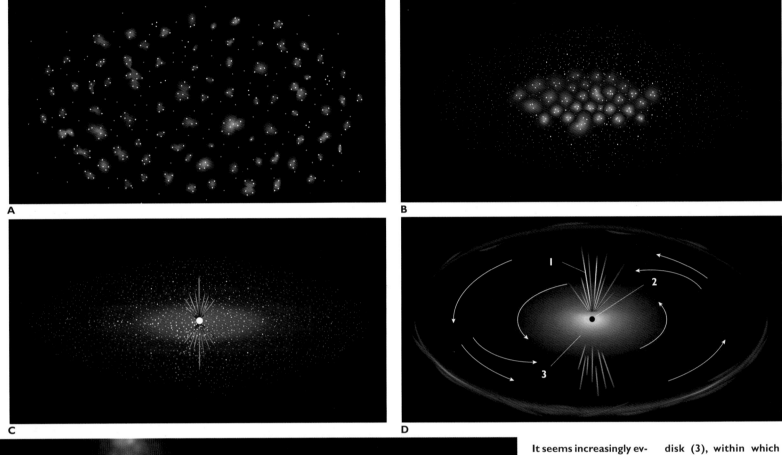

A

B

C

D

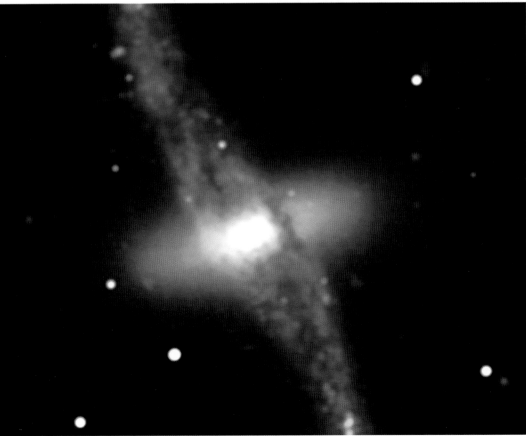

It seems increasingly evident that the light produced by many extragalactic objects is derived from phenomena more powerful than the nuclear reactions that take place in the stars of which they are composed. The brightest among these are quasars, but other celestial objects have nuclei that are shaken by energy production. It is believed that normal galaxies, "exotic" objects and quasars are linked in an evolutionary chain, as shown above. When a galaxy is formed from a large cloud of gas that contracts through its own gravity (A, B), the central regions undergo a collapse (C), contracting into a supermassive black hole (D). This sucks in the gas in the nucleus, which arranges itself into an accretion disk (3), within which it accelerates to speeds close to that of light. When the gas falls into the black hole (2), it emits a very intense light (1). Small black holes feed normal galaxies; medium and large black holes feed active galactic nuclei and quasars.

At left, one of the earliest images produced, in May 1998, by the first of what will eventually be four **ESO VLT** telescope units. NGC 4650A is a peculiar galaxy in which two components different stellar populations: an S0 galaxy composed of old, reddish stars, and a ring perpendicular to it made up of young blue stars, dust and gas.

■ Composite of seven images taken with blue, green and red filters for a total exposure time of 64 minutes.

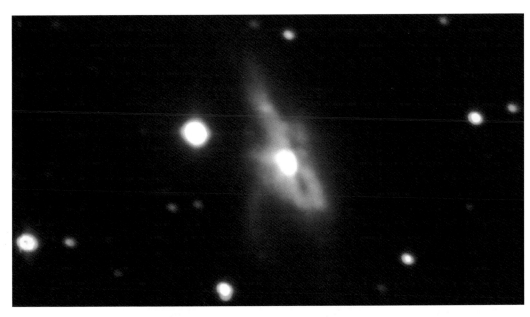

Left, the "starburst" galaxy **NGC 6240** in Ophiuchus, 300 million light-years away. Objects of this type are extremely luminous because of the thermonuclear energy derived from the simultaneous formation of an enormous quantity of stars in their nuclei, probably triggered, as we have noted, by galactic collisions. It is easy to see the bright nucleus and the peculiar outer structure. NGC 6240 is also a mega-maser ("maser" stands for microwave amplification by stimulated emission of radiation), a galaxy that emits radio waves of considerable intensity through stimulated emission from clouds of molecules and interstellar radicals (such as water, hydroxyl radicals and formaldehyde) that surround a central radio source.

■ Computer composite of three **CCD** images by Rudolph Schild.

In the 1970s, Benjamin Markarian, at the Byurakan observatory in Armenia, discovered a few galaxies with powerful emissions in blue and ultraviolet light, a sure sign of the presence of many very young stars, probably produced by tidal interactions between galaxies. Far right, Markarian 496 in Draco shows a double nucleus characterized by intense star-forming activity and two long tails of tidal origin.

■ False-color CCD image created by J.M. Mazzarella and T.A. Boroson from images taken in blue and yellow light with the 1.3-m McGraw-Hill Telescope of the KPNO.

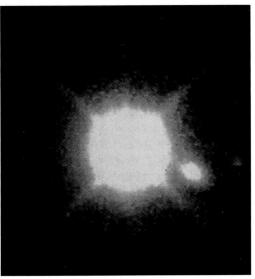

Above, PKS 2155-30, by far the most luminous BL Lac type object known. BL Lacs are highly luminous, compact radio sources. Their brightness is greatly variable, and they are found at the center of giant elliptical galaxies.

■ Composite picture produced from various CCD images, with a total exposure time of 5 minutes, taken with the ESO's New Technology Telescope (NTT).

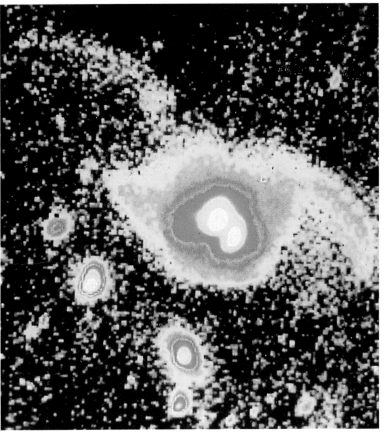

*Active Galactic Nuclei (AGN) are objects in which the production of energy occurs (drawing on page 110) at the expense of a central supermassive black hole that feeds on the gas that falls into it from the nucleus of the mother galaxy or nearby interacting galaxies. AGNs comprise quasars, radio galaxies, Seyfert galaxies and BL Lac objects. Their history dates from the 1940s, when Carl Seyfert, a student of Humason and Minkow-sky at Mount Wilson Observatory, discovered spirals with a nucleus that had a starlike appearance, rich in hot, young stars and characterized by explosive activity. In the 1950s and 1960s, elliptical galaxies were discovered with bright, variable and strongly radio-emissive nuclei. In 1968, it was also discovered that an object that had been considered a variable star, BL Lacertae, exhibited clear similarities to quasars, radio galaxies and Seyfert galaxies. Currently, the opinion is that all these objects have a similar nature. Their different appearances result mainly from the orientation of the accretion disk and the corresponding jets, with respect to our line of sight. The jets are produced by in-falling gas which arranges itself into two gigantic vortices—labeled "1" in the drawing on page 110—protruding from the poles of the disk's axis of rotation.*

Below, the nucleus of M51 photographed by the Hubble Space Telescope: the X-shaped structure in the center could be the signature of the accretion disk of a black hole of 1 million solar masses.

■ Picture taken with the PC-1 in green light by Holland Ford (Johns Hopkins University).

Right, the Seyfert galaxy NGC 1566, the brightest member of a group of galaxies in the southern constellation Dorado, is about 50 million light-years away.
Image enhancement techniques used to produce this color picture enable us to see the spiral arms stretching to a size that is almost four times what is ordinarily visible.
■ Tricolor composite created by David Malin from three plates exposed by Bill Pence on the AAT for 25, 25 and 35 minutes.

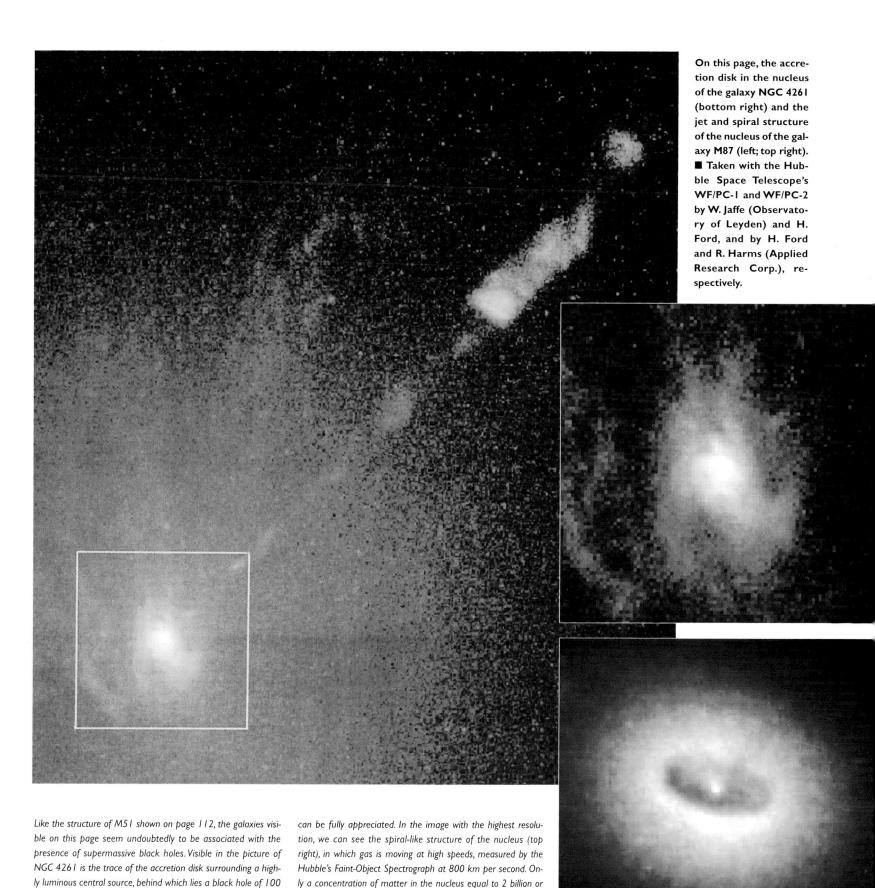

On this page, the accretion disk in the nucleus of the galaxy **NGC 4261** (bottom right) and the jet and spiral structure of the nucleus of the galaxy **M87** (left; top right). ■ Taken with the Hubble Space Telescope's **WF/PC-1** and **WF/PC-2** by **W. Jaffe** (Observatory of Leyden) and **H. Ford**, and by **H. Ford** and **R. Harms** (Applied Research Corp.), respectively.

Like the structure of M51 shown on page 112, the galaxies visible on this page seem undoubtedly to be associated with the presence of supermassive black holes. Visible in the picture of NGC 4261 is the trace of the accretion disk surrounding a highly luminous central source, behind which lies a black hole of 100 million solar masses. In the wide-field picture of M87 (above), the great corresponding jet coming out of the galaxy's nucleus can be fully appreciated. In the image with the highest resolution, we can see the spiral-like structure of the nucleus (top right), in which gas is moving at high speeds, measured by the Hubble's Faint-Object Spectrograph at 800 km per second. Only a concentration of matter in the nucleus equal to 2 billion or 3 billion solar masses, that is to say, a gigantic black hole, could prevent gas of such velocity from escaping.

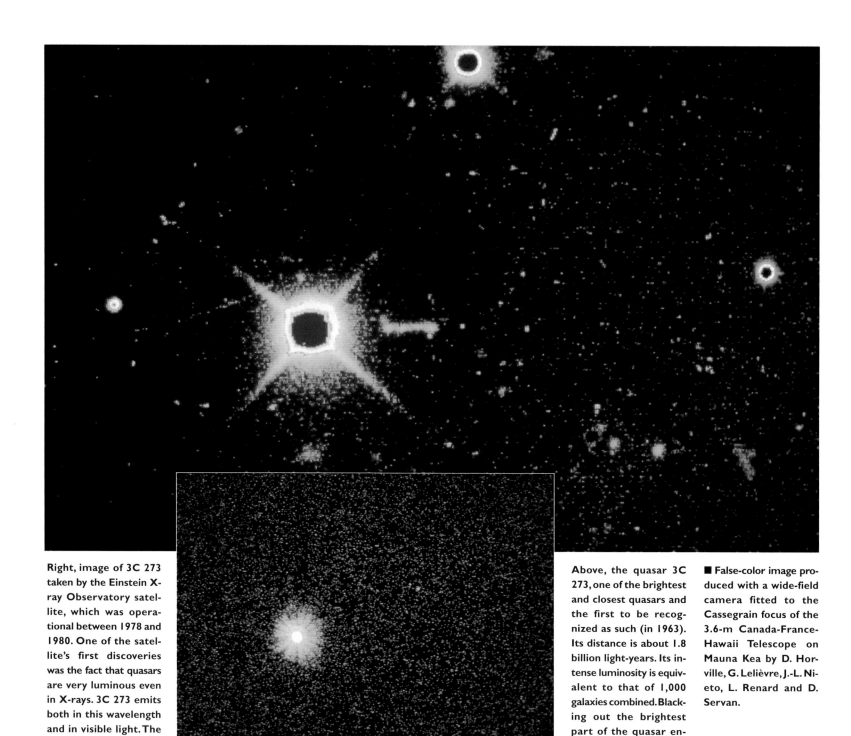

Right, image of 3C 273 taken by the Einstein X-ray Observatory satellite, which was operational between 1978 and 1980. One of the satellite's first discoveries was the fact that quasars are very luminous even in X-rays. 3C 273 emits both in this wavelength and in visible light. The emission is predominantly concentrated at the nucleus.

Above, the quasar 3C 273, one of the brightest and closest quasars and the first to be recognized as such (in 1963). Its distance is about 1.8 billion light-years. Its intense luminosity is equivalent to that of 1,000 galaxies combined. Blacking out the brightest part of the quasar enables us to see the jet of 3C 273.

■ False-color image produced with a wide-field camera fitted to the Cassegrain focus of the 3.6-m Canada-France-Hawaii Telescope on Mauna Kea by D. Horville, G. Lelièvre, J.-L. Nieto, L. Renard and D. Servan.

The history of the discovery of quasars is one of the strangest in modern astronomy. During the 1960s, attempts were made to identify in visible light a few celestial objects that emitted radio waves. The radio source 3C 48 (3C stands for "third catalog of radio sources of the Cambridge Observatory") was identified with a tiny star of just 16th magnitude which, given the power of the source, appeared to be very dim. In 1960, American astronomer Allan Sandage obtained a spectrum of it that showed anomalously spread-out lines. In 1963, Dutch astronomer Maarten Schmidt obtained a spectrum of another optical counterpart of a radio source, 3C 273, a tiny 13th magnitude star, which showed a substantial resemblance to that of 3C 48. Schmidt attempted to prove a hypothesis that the spectral lines appeared spread out because of a very strong redshift produced by an object that was rapidly moving away. It worked, and the lines thus came to correspond to those already known in other celestial objects. The procedure was repeated for 3C 48 and other similar sources, for which the term quasi-stellar radio sources (QSRS), or quasars, was coined.

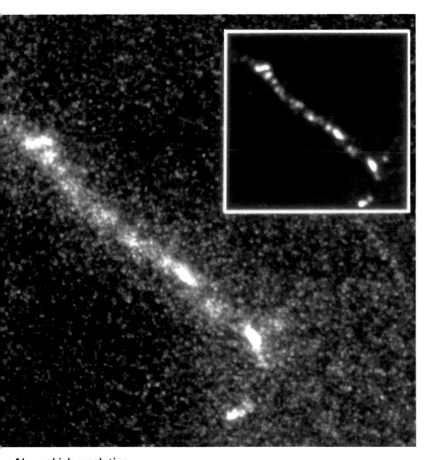

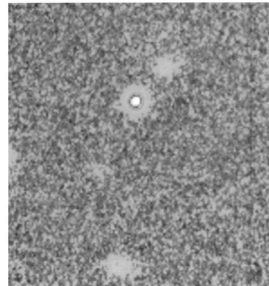

Above, high-resolution image of the jet of 3C 273 taken by the FOC of the Hubble Space Telescope. Further enhancement of the picture (see inset) reveals the characteristic structure of globules and knots, the result of the interaction between the jet and a shell of expanding matter from the quasar.

■ Composite of 18 22-minute exposures taken with a blue filter and three polarizers by Robert Thomson, Craig Mackay and Alan Wright.

Below right, the jet of the galaxy 3C 66B is extremely complex. Its outer end seems unraveled and appears to be literally made up of two intertwined elements, in a structure like that of hemp rope. In the picture, to better show the jet, the light of the galaxy has been removed. The jet is about 10,000 light-years long, double the length of M87's jet, but only one-fifteenth the length of the jet of 3C 273.

■ Photographed in ultraviolet light with the Faint Object Camera of the Hubble Space Telescope by Duccio Macchetto and other members of the ESA's FOC team.

Below left, the most distant quasar known, PC 1247+3406 in Canes Venatici, photographed with the largest optical telescope in the world, the William Keck Telescope. The Keck has a diameter of no less than 9.8 m and is located at a height of 4,150 m, at the top of the extinct volcano Mauna Kea, in Hawaii.

The quasar is just above the center of the picture. Its light started traveling when the universe was just a billion years old. The other luminous spots are extremely distant galaxies.

■ Photographed in infrared light, at a wavelength of 2.2 microns, with the Near-Infrared Camera (a composite of about 100 37-minute exposures).

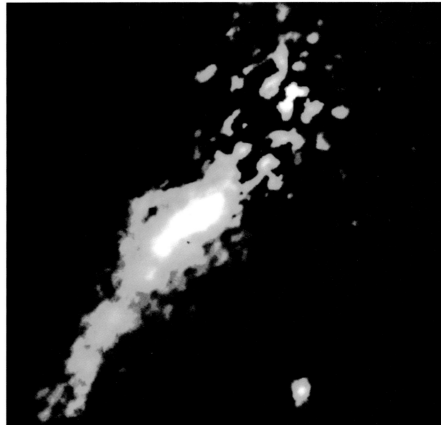

Not all quasars emit radio waves. Indeed, most objects of very high redshift, having a starlike appearance and showing a high energy emission—all the characteristic traits of quasars—are radio-quiet. For them, the quasar acronym (QSRS, for "quasi-stellar radio source") is inappropriate. It was therefore decided to call such objects, whether radio-emissive or radio-quiet, simply QSOs (quasi-stellar objects) or, as Maarten Schmidt suggested in 1970 (a view endorsed by the present author), simply quasar, without the final "s." It was quickly understood that quasars could be anything except stars. If their redshifts were related to the expansion of the universe, then they had to be very distant. Indeed, the greater the redshift, the greater the velocity of recession and, according to Hubble's law (see page 129), the greater the object's distance, in billions of light-years. How could ordinary stars be visible at such a distance, where not even galaxies could be seen? Simple calculations made it evident that a quasar had to be 10 to 1,000 times brighter than a galaxy. Today, there is scarcely any question that a quasar is the active nucleus of a galaxy which is too distant to be visible.

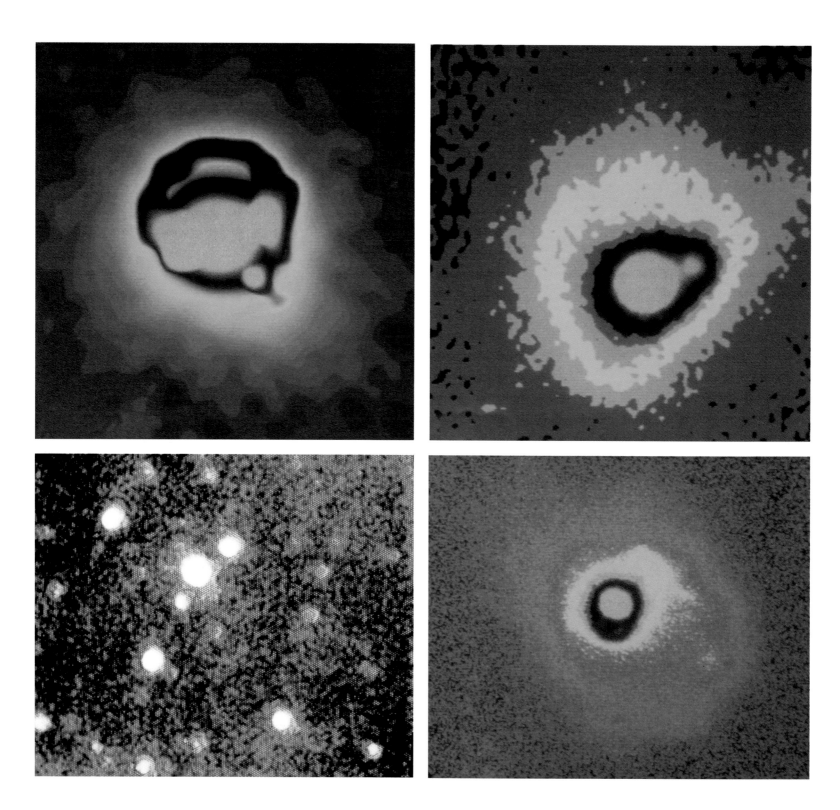

Computer composite images made by Nigel Sharp at the KPNO, showing three quasars: IRAS 00275-2859 (top left), PG 1613 (top right) and Mk 231 (bottom right). The picture at bottom left, showing the quasar 3C 275.1, the first to be found at the center of a galaxy cluster, is a pseudocolor image taken by Paul Hintzen and John Stocke with the 3.8-m Mayall Telescope of the KPNO. Five billion light-years away, the quasar lies within a rotating cloud of gas that may be an elliptical galaxy.

*The three quasars could represent separate stages in the life of a quasar, whose formation, as has been seen, was favored by interactions between galaxies. IRAS 00275-2859 is about to collide with a second nucleus. PG 1613 shows a collision and the deformation of the surrounding galaxy. In Mk 231, finally, the collision appears complete, forming (through gravitational interaction) tidal tails and filamentary clouds.*

Right, the quasar Markarian 205, in the constellation Draco, appears to lie beside the nucleus of the galaxy NGC 4319. However, the galaxy's redshift is very minor; it is 69 million light-years from us, whereas the quasar is believed to be 850 million light-years away. On the other hand, a bridge of matter joining the two objects is clearly visible.

■ This image was taken using a blue filter and 10 minutes' exposure with a CCD camera fitted onto the 3.8-m Mayall Telescope of the KPNO. Subsequent false-color enhancement reveals the subtle differences in luminosity within the field shown.

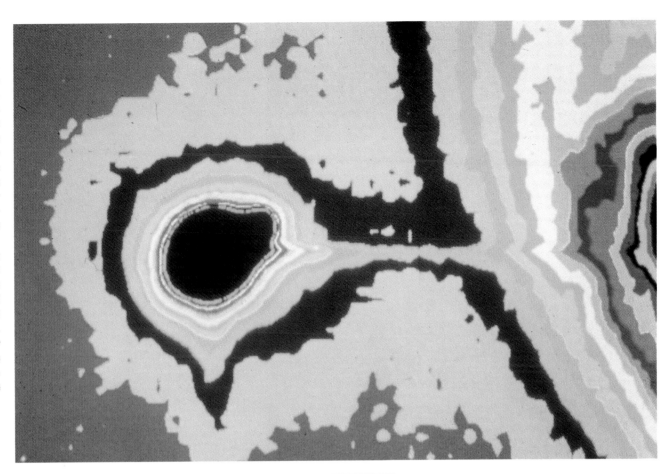

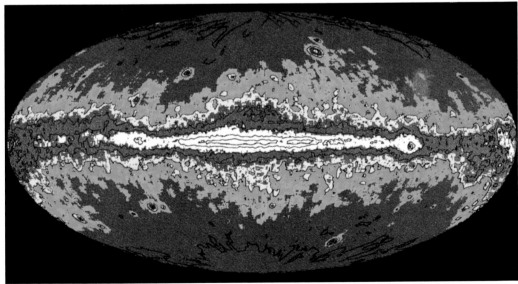

Left, in this gamma ray map of the sky created by **NASA's** orbiting **Gamma Ray Observatory (GRO)**, launched in 1991, a few quasars stand out, remaining faithful to their role as the brightest objects in the universe even at this wavelength. Each one shows gamma ray emissions about 10,000 times greater than those of the Milky Way. The horizontal swath that crosses the map is produced by the diffuse emissions of our own galaxy.

■ This false-color image is the product of a year's worth of observations with the **EGRET** (Energetic Gamma Ray Experiment Telescope) onboard **GRO**.

The presence of discordant redshifts between extragalactic objects close to one another and linked by bridges of matter, real or apparent, is not unique to the case shown above but extends to many other examples, documented by American astronomer Halton Arp. This work, considered unorthodox and almost heretical, sadly put Arp on the fringes of the astronomical community for many years. However, even a conventional enquiry cannot deny the existence of a few bridges of matter, and the possibility of a pure perspective alignment between the objects has to be rejected. Statistically, the number of alignments between objects with different redshifts seems more than a casual distribution might lead to. Even the hypothesis of a distorting influence due to gravitational lensing phenomena, which would facilitate the apparent perspective alignment between sources far and near, does not seem workable. It has to be admitted that not all quasars' redshifts are cosmological in origin (that is, due to the great recessional velocity of objects participating in the expansion of the universe). An alternative hypothesis put forward by William Saslaw is that some quasars are compact objects violently ejected from the nucleus of a galaxy following explosive phenomena.

# The Hubble Space Telescope.

The first feasibility study by a group of NASA scientists looking into the idea of a space telescope dates from 1962.

The possibility of observing from beyond the Earth's atmosphere was considered to be a priceless advantage. Only in those conditions would it be possible to exploit fully, both in terms of resolution and light-gathering capacity, the power of an astronomical telescope.

Following this, other study groups in 1965, 1969 and 1972 produced their findings, from which it emerged that the construction and launch of a space telescope was not only desirable but also realistic. In 1973, NASA put a team of scientists in charge of establishing such a project, and in 1977, after various hesitations and uncertainties, NASA assigned a new team to the final planning and production, which was ratified that same year by Congress.

The telescope was due to be launched in 1983, but through a series of technical and logistical problems, this was postponed to 1985, then to 1986. In that year, however, the dreadful Challenger shuttle catastrophe occurred, when the space shuttle exploded soon after takeoff, with seven astronauts onboard, and this blocked the USA's entire space program for 3 years. The cost, too, initially estimated at $500 million, escalated to three times that amount by the time the launch took place.

Finally, in April 1990, the Hubble Space Telescope was launched.

Mounted on a 13-m-long satellite, 4.3 m wide and 10.5 tonnes in weight, the telescope has a primary mirror 2.4 m in diameter. Light enters through the opening in the telescope (see figure below) and is reflected from the primary to a secondary mirror, placed at a distance of almost 5 m, which reflects it a second time through a hole in the center of the primary mirror. The image is focused several meters behind the primary mirror, where the five scientific instruments are placed. Four are mounted along the optical axis of the instrument. There are two spectrographs—instruments for analyzing the light of celestial objects—one to study

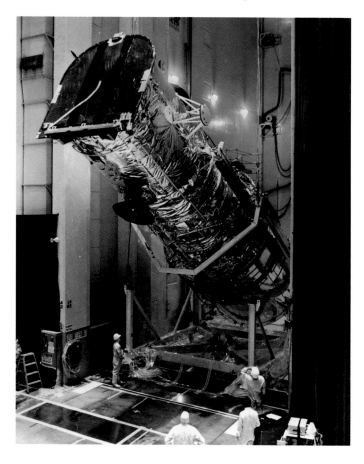

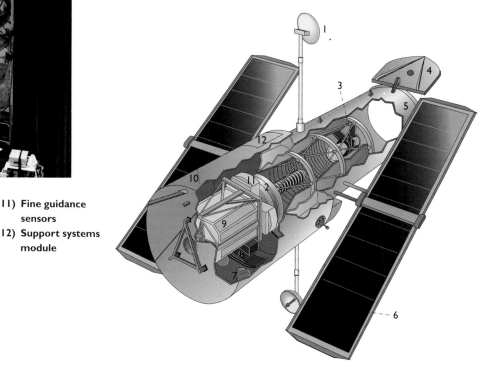

Above, the Hubble Space Telescope still in its protective wrappings prior to launch and, at right, a cross-section diagram of the instrument:

1) Radio antenna
2) Primary mirror
3) Secondary mirror
4) Aperture door (open)
5) Telescope tube
6) Solar panels
7) Star-tracking device
8) Wide Field/ Planetary Camera
9) Axial scientific instrument module
10) Rear panel
11) Fine guidance sensors
12) Support systems module

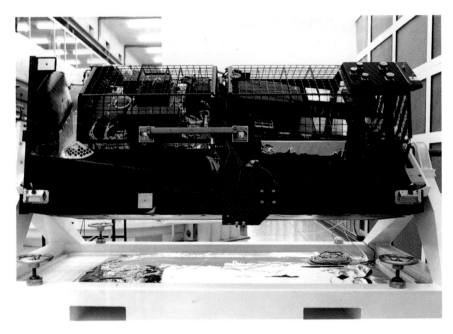

ever, the five instruments were designed to cover a range of "only" 115 to 1,100 nm.

The Hubble Space Telescope was named after Edwin Hubble, the greatest astronomer of the 20th century. He was the discoverer of, among other things, the expansion of the universe and the extragalactic nature of spiral nebulas, and the inventor of the first classification system for galaxies.

**On this page (left), an exterior view of the Faint Object Camera. Below, cross-section diagram of the Wide Field/Planetary Camera:**

1) Heat converter
2) External radiator
3) Pyramid-shaped mirror
4) Incident light path
5) Deflecting mirror
6) Filter-bearing reel
7) Diaphragm
8) Tube for directing the flow of ultraviolet light
9) Ultraviolet light from the Sun
10) CCD camera

very faint objects and the other to analyze brighter sources, but with a higher resolution. There is also a High-Speed Photometer, an instrument for measuring with extreme precision the intensity of the light coming from astronomical sources, plus a Faint Object Camera (FOC). The FOC (below), built by the European Space Agency (ESA), is the instrument able to make the most use of the Space Telescope's maximum resolution, down to 0.07 arc-second, at least 30 times better than the best terrestrial telescopes—rather like seeing a coin clearly at 700 km distance.

The FOC has two image intensifiers, which are roughly similar to the cathode ray tube in a videocamera. Mounted perpendicularly to the optical axis is

the Wide Field/Planetary Camera (WF/PC). The WF/PC (right) works in two complementary ways, thanks to the presence of a small, adjustable pyramid-shaped mirror able to direct the incident light at the appropriate angle. It can pick up images of celestial objects in wide fields at low magnification or in narrower fields but with greater magnification—for example, for photographing the planets.

Altogether, by exploiting the absence of atmosphere, the telescope is designed to respond to a very wide spectral range, able to detect light with wavelengths ranging from 115 nm (far ultraviolet) to 1 cm (far infrared), a range about 3,000 times greater than in terrestrial telescopes. Initially, how-

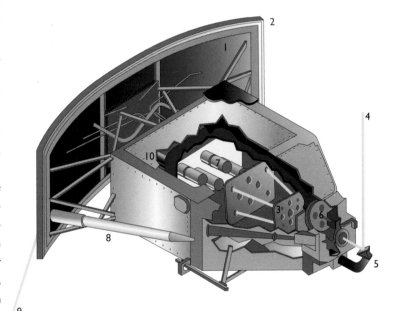

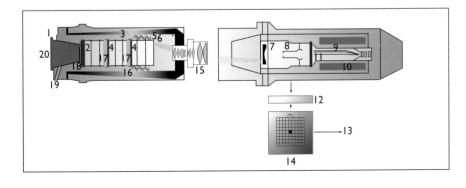

| Left, cross-section diagram of the Faint Object Camera: | |
|---|---|
| 1) Diaphragm | 10) Scanning coil |
| 2) Photocathode | 11) Silicon wheel |
| 3) Magnetic focusing device | 12) Pre-amplifier |
| 4) Photocathodes | 13) Data storage unit |
| 5) Coil | 14) Video enhancing unit |
| 6) Phosphorus emission | 15) Collimation lenses |
| 7) Photocathode | 16) Three-stage intensifier |
| 8) Television tube | 17) Phosphorus |
| 9) Red light | 18) Frontal plate |
| | 19) High-voltage conductor |
| | 20) Incident light |

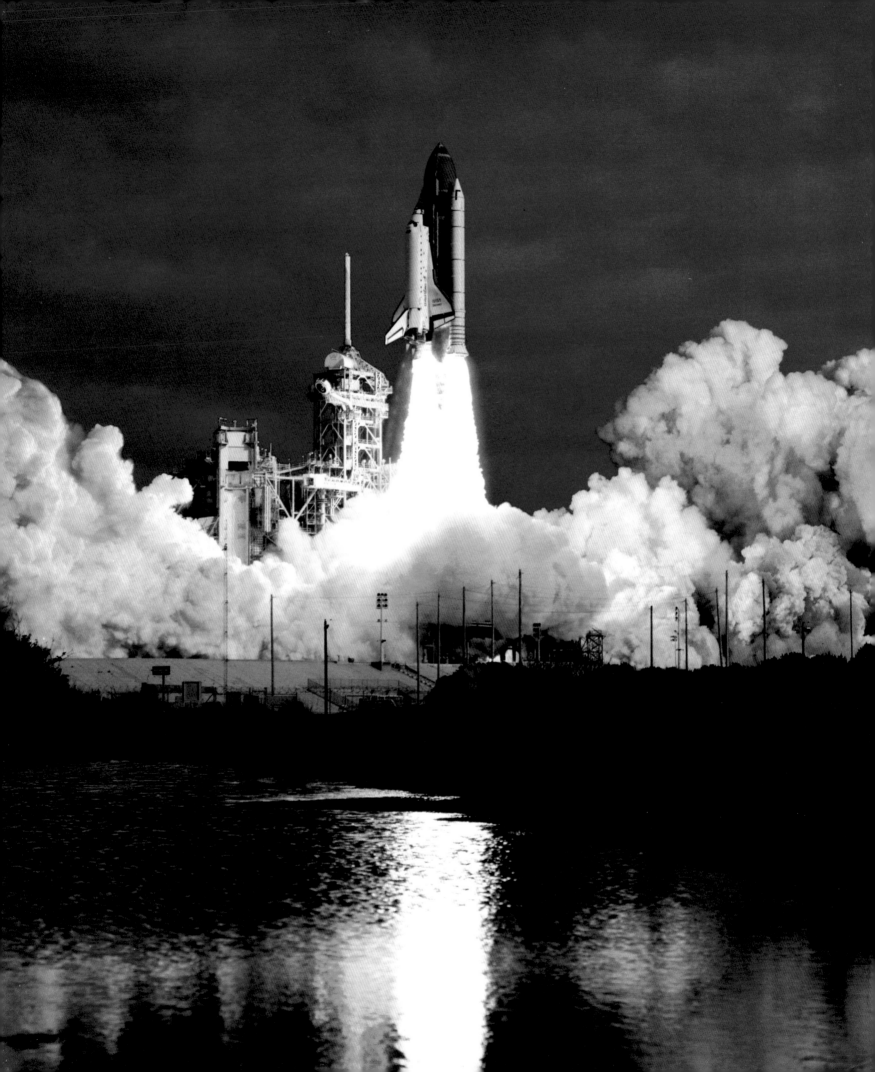

Facing page, an important moment in astronomy: on April 24, 1990, after a long wait and many postponements, the era of the Space Telescope finally began. Onboard the space shuttle Discovery, the telescope left Cape Canaveral in Florida, amid colorful flames and a rumble of thunder. The departure of a space shuttle is in itself a fascinating and unique spectacle, but the trepidation that accompanied this particular launch increased tension and emotions out of all proportion: the jewel in the crown of the international astronomical research community was beginning its journey, and with it went the hopes of thousands of researchers and the dreams and wishes of millions of enthusiasts. What would the Hubble discover? Would it be able to discover planets of other stars? Would it see faint Cepheids in distant galaxies, contributing to an improved knowledge of the size of the universe? Would it be able to find black holes at the center of globular clusters and active galaxies? Would it resolve definitively the enigma of quasars? Would it actually photograph stars in the process of forming? What views would it obtain of distant galaxy clusters on the outer edges of the universe?

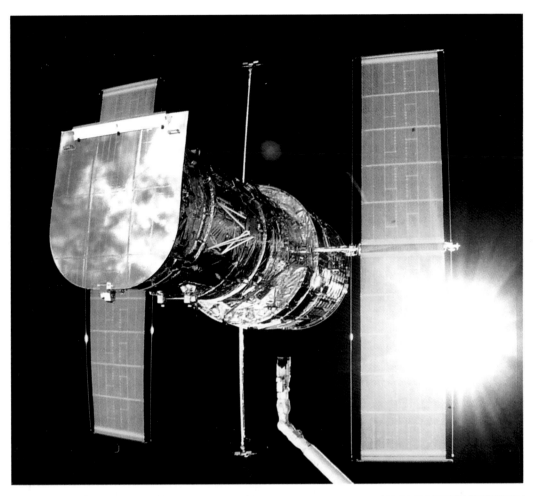

The moment when, on April 25, 1990, the Hubble Space Telescope (HST) was released from the mechanical arm (visible below it) of the shuttle. The orbit assigned to the instrument, at a height of 614 km, is higher than originally anticipated. The extra height was needed because the launch took place during a period of considerable solar activity, the effect of which was to make the Earth's atmosphere expand, thereby increasing the problem of atmospheric friction that would lead to premature lowering of the satellite's orbit.

■ Both photographs were taken with a 70-mm IMAX camera fitted onto the shuttle's loading bay.

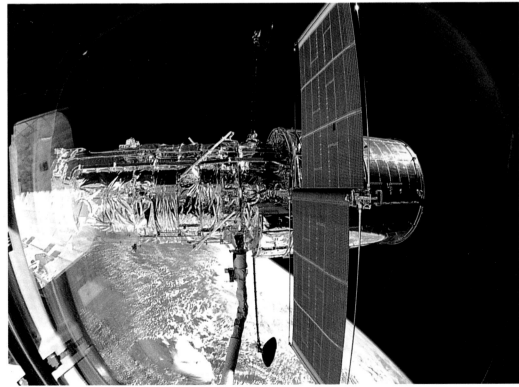

Prior to its release, the Hubble Space Telescope's solar panels were extended. These gather the energy that makes the various systems onboard function. The Canadarm (remote manipulator system), visible below the telescope in the photograph, maneuvered by mission specialist Steven Hawley (from inside the shuttle), unfurled the first of the two panels, which is shown at about three-fifths of its full extension. The solar panels, each 12 m long (almost as long as the telescope), were built by the European Space Agency.

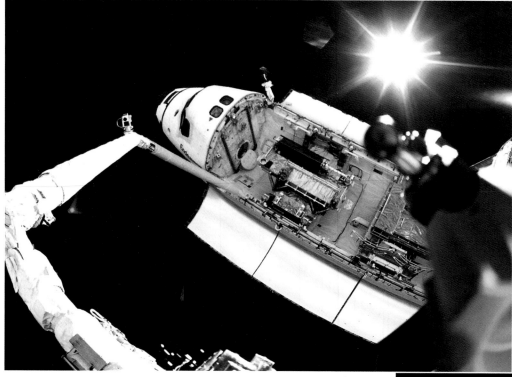

Left, a picture of the space shuttle Endeavor's loading bay, taken by an astronaut near the top of the telescope during the Hubble repair mission. Visible in the foreground is the shuttle's mechanical Canadarm, an indispensable tool for the astronauts' work in space, maneuvered during the mission by ESA astronaut Claude Nicollier.

In the black background of the sky, the Sun.

On these pages, pictures of the historic mission, considered to be the most important and complex to be carried out to date by human beings in space, which took place from December 2 to 13, 1993, and led to the repair of the Space Telescope. The difficulties were so great that many astronomers hoped that, after the repairs, the instrument would at least function as well as before. A maintenance mission had been anticipated, but of a more ordinary nature. Now, however, the optics had to be corrected, the high-resolution spectrograph had to be reset, and a whole series of malfunctioning systems had to be repaired or replaced, including the solar panels and three of the six gyroscopes onboard (devices that keep the telescope stable in space). A correcting apparatus, called COSTAR, had been designed for the axial instruments. The WF/PC was totally replaced (as had been anticipated) by a decidedly more sophisticated model, the WF/PC-2, containing corrective lenses in its interior. The WF/PC-2 is fitted with more powerful CCD sensors, particularly more sensitive to ultraviolet light. At right, astronaut Jeffrey Hoffman is taking away the WF/PC-1 after installing the WF/PC-2.

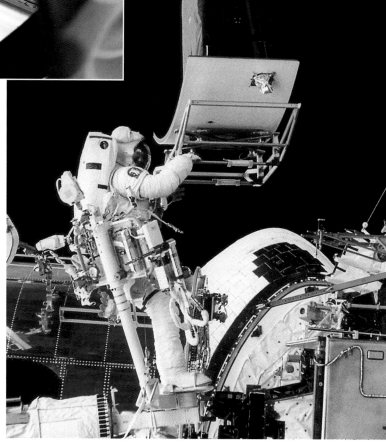

At the end of June 1990, after 2 months of adjustments, tests and optical collimations, NASA released a statement that had the effect of a cold shower: the Hubble Space Telescope was not able to focus properly on stellar images. The cause, which was soon realized, lay in a serious fault in the primary mirror, a high degree of spherical aberration that had not been picked up by the insufficiently rigorous (as was later found out) tests carried out before the launch. Because of this, the Hubble, instead of seeing the stars as pinpoints, was seeing them as large, out-of-focus patches. This translated into a considerable lessening of its powers in terms of both resolution (the capacity to see small details) and light-gathering capacity (the ability to see very faint stars). The Space Telescope was faring no better than a terrestrial one. After an initial moment of surprise and indignation, it was realized that the problem could be remedied. In the short term, this involved—at least as far as the brighter objects were concerned—computer correction of the images, intended to eliminate the irksome bright halos that were interfering with image quality. Looking further ahead, a repair mission was planned that would provide the telescope with corrective optics.

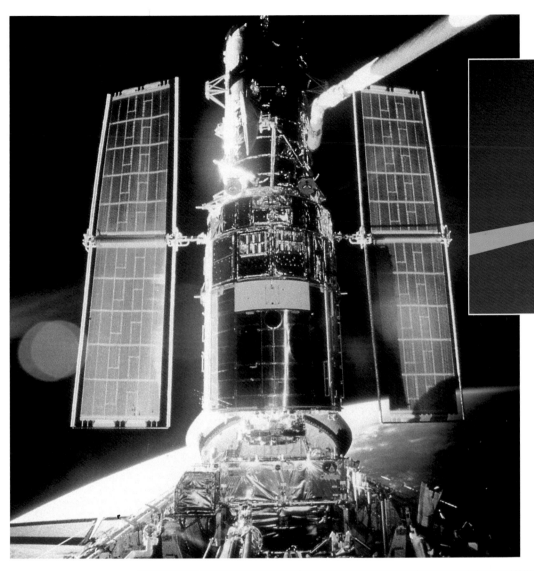

At left, the Hubble is about to be released from the shuttle's Canadarm after an intensive 5 days of repair work, arranged in 8-hour shifts. The other astronauts engaged in this undertaking were Story Musgrave (mission specialist, together with Thornton, Akers and Hoffman), Richard O. Covey (commander) and Kenneth Bowersox (shuttle pilot).

Above, a drawing showing the **COSTAR** (Corrective Optics Space Telescope Axial Replacement). A robotic arm, containing three pairs of mirrors, projects onto the optical axis of the telescope. Each pair is made up of a reflecting mirror and a correcting mirror and is able to correct the incident light for each of the axial instruments. A beam of light is intercepted by the first mirror and bounced backward, where it is picked up by the second, correcting mirror, on the outstretched "fingers" of the robotic arm, which ensures it is sent to the instrument being used.

1) **Optical bank**
2) **Mirror-bearing "fingers"**
3) **Light beam**
4) **Reflecting mirror**
5) **Correcting mirror**

On the right, astronaut Kathryn Thornton taking the **COSTAR** container (the same size as a refrigerator and weighing 290 kg) from the loading bay of Endeavor, while her companion Thomas Akers prepares to assist her in installing the instrument. So as not to further complicate the repair maneuvers, **COSTAR** replaced the High-Speed Photometer, which thus had to be sacrificed.

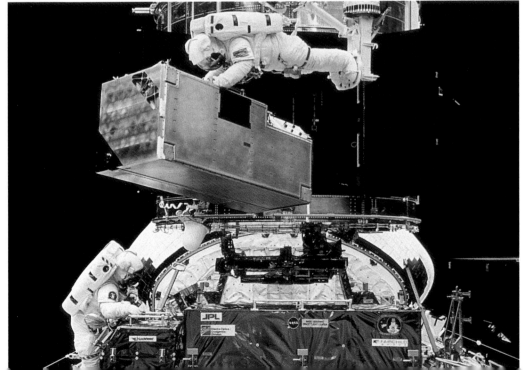

The capabilities of the **WF/PC-2** are spectacularly demonstrated in this March 1997 image of Mars, the best ever obtained by the Space Telescope. The resolution is 0.045 arc-second, showing features as small as 22 km in diameter on the Martian surface. Clearly visible is the north polar cap, which is reduced to plain ice at the beginning of the Martian summer. Similar images were used to help prepare for the Mars Pathfinder and Mars Global Surveyor missions (see page 24).

■ Tricolor composite of three images taken with blue (433 nm), green (554 nm) and red (763 nm) filters.

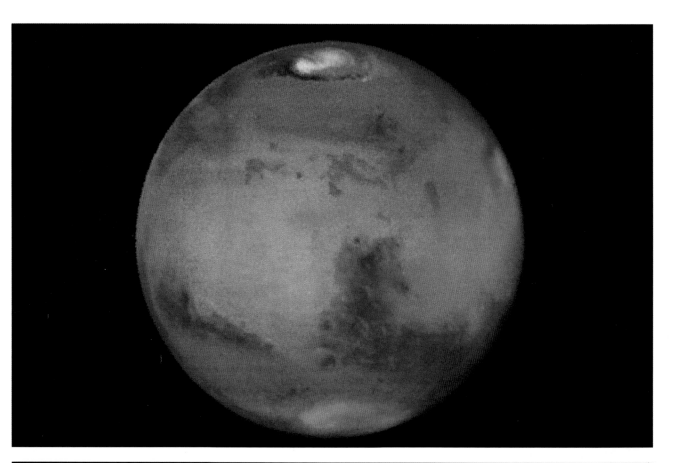

The planet Jupiter photographed on February 13, 1995. Hubble has captured planetary images comparable in richness of detail to those from the Viking and Voyager missions to Mars, Jupiter and Saturn. These illustrations leave little doubt as to its prowess (see also pages 31, 34 and 127). Compiled by the Hubble Space Telescope team led by Reta Beebe of the University of New Mexico, this image is part of a series of shots that document rapid changes in Jupiter's atmosphere.

■ James Westphal (California Institute of Technology), the designer of the **WF/PC**, produced this picture, which is part of a wider program of study of the atmosphere of Jupiter, by combining three separate shots using blue, green and red filters.

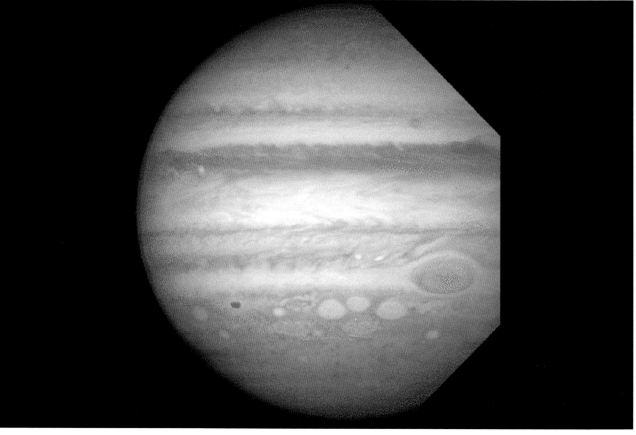

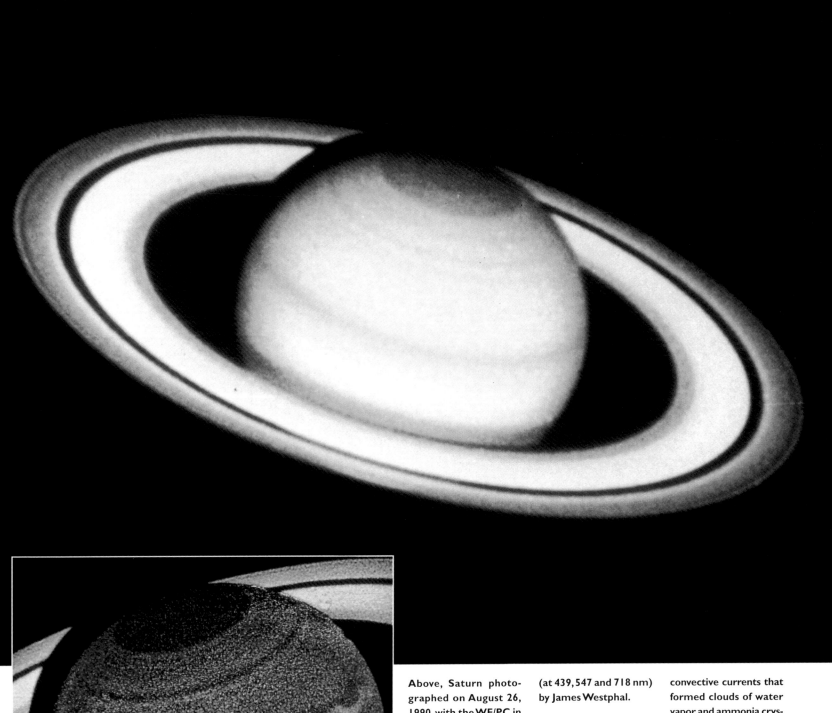

Above, Saturn photographed on August 26, 1990, with the WF/PC in wide-field mode. The picture reveals for the first time atmospheric details of the north polar zone, which was not explored during the two Voyager missions.
■ Composite of three images photographed in blue, green and red light (at 439, 547 and 718 nm) by James Westphal.

At left, an unusual view of Saturn, taken by Westphal with the WF/PC in planetary mode on November 9, 1990, while an extended atmospheric disturbance was traversing the disk of the planet. The heat generated produced powerful convective currents that formed clouds of water vapor and ammonia crystals which then expanded disproportionately under the action of Saturn's very strong winds.
■ Composite of two images taken in blue light and infrared.

Below, Comet Hale-Bopp photographed on eight different dates beginning in September 1995. The shots document changes in its nuclear region as it approached the Sun. False-color-enhanced images taken by Harold Heaver of Johns Hopkins University using the WF/PC-2.

At right, Uranus photographed in July 1997 by Hubble's Near Infrared Camera and Multi-Object Spectrometer (NICMOS, see page 133). When Voyager 2 reached Uranus (see page 32), it saw a featureless atmosphere, but Hubble has shown evidence of various

atmospheric features on several occasions. Six can be seen clearly in these two images. Also clearly seen is the planet's system of rings, which are highly visible in infrared, plus eight of its satellites.

■ Composites of three infrared shots taken at 1.1, 1.6 and 1.9 microns by Erich Karkoschka of the University of Arizona.

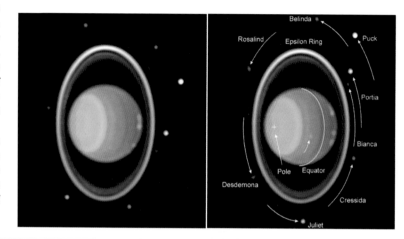

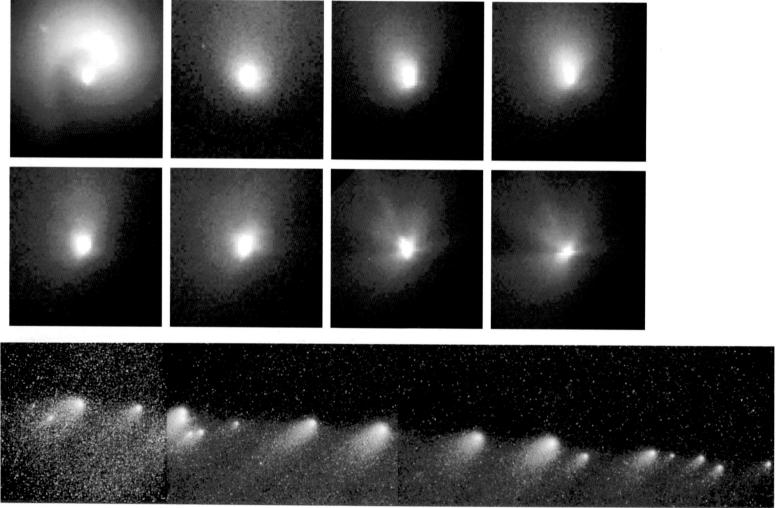

On March 15, 1993, Eugene and Carolyn Shoemaker and David Levy discovered a new and highly unusual comet with the 46-cm Schmidt telescope on Mount Palomar. Instead of being a single diffuse patch, the comet looked like a kind of bar surrounded by an elongated halo. Subsequent shots with more powerful telescopes revealed its true nature: a string of more than 20 comet fragments near the planet Jupiter. Initially orbit-

ing around the Sun, Comet Shoemaker-Levy 9 was captured by Jupiter in 1970, and on July 8, 1992, a suicide close approach to the giant planet brought about its rupture. This event further modified its orbit, so much so that the comet fragments would fall onto Jupiter from July 16 to 22, 1994. This was the first time that humanity would witness such an event at first hand, and immediately, speculation was rife as to what would actually be

seen. Because the impacts occurred on the opposite side of the planet to the one visible from Earth, they would never be observed directly. But their effects would be seen in the form of extensive atmospheric disturbances that would become visible once Jupiter had rotated slightly toward us, exhibiting the scars of the impacts. The Hubble Space Telescope, once again, produced the best pictures of the event.

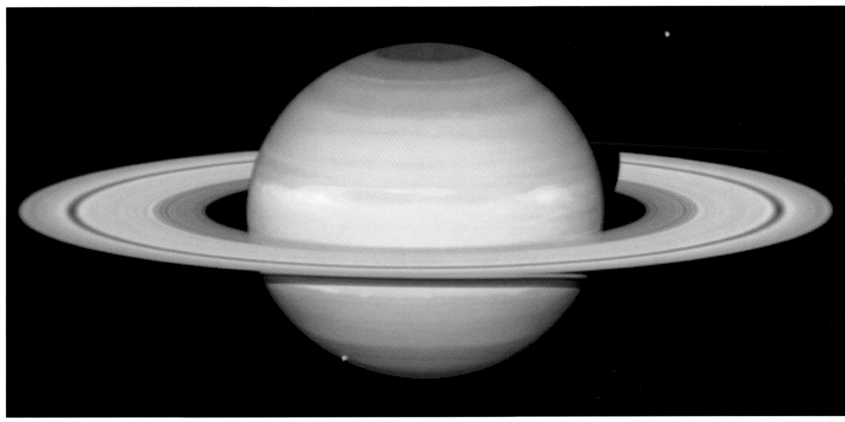

Above, another planetary image taken by **NICMOS**, this time of Saturn. False-color image produced from an infrared shot taken on January 4, 1998, by Erich Karkoschka.

At right, a panoramic view of eight scars left by the impacts of fragments of Comet Shoemaker-Levy 9. The effect of the collisions was far greater than had been anticipated; the dark spots produced were visible even in small telescopes. The energy released was equal to 1,000 times that of the Earth's entire nuclear arsenal.

■ Tricolor image by the HST Comet Team from Planetary Camera 2 photographs in blue, green and red light at 410, 555 and 953 nm.

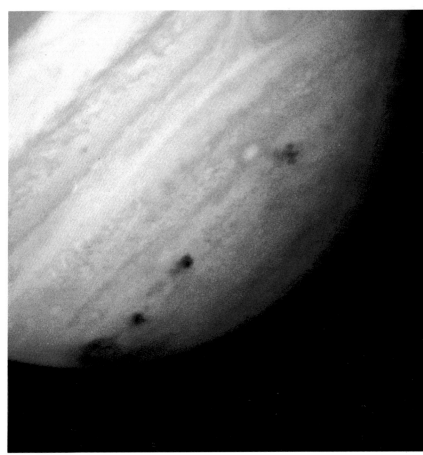

A high-resolution Hubble image of a portion of the inside lip of the Helix Nebula (see page 64) reveals thousands of globular objects, comet-like in appearance. Though they appear minuscule in the image, each of these globules is enormous: their heads are at least twice the size of our solar system, while the tails are about 1,000 astronomical units long—1,000 times the Earth-Sun distance. It is thought that the globules formed because hot low-density gas emitted by the nebula's central star collides with the surrounding colder, denser gas emitted 10,000 years earlier. The force of the collision causes the cloud surrounding the star to fragment into smaller, denser globules resembling falling paint drops.

■ Image composed by Robert O'Dell and Kerry P. Handron of Rice University from shots taken in red, green and blue light with the WF/PC-2.

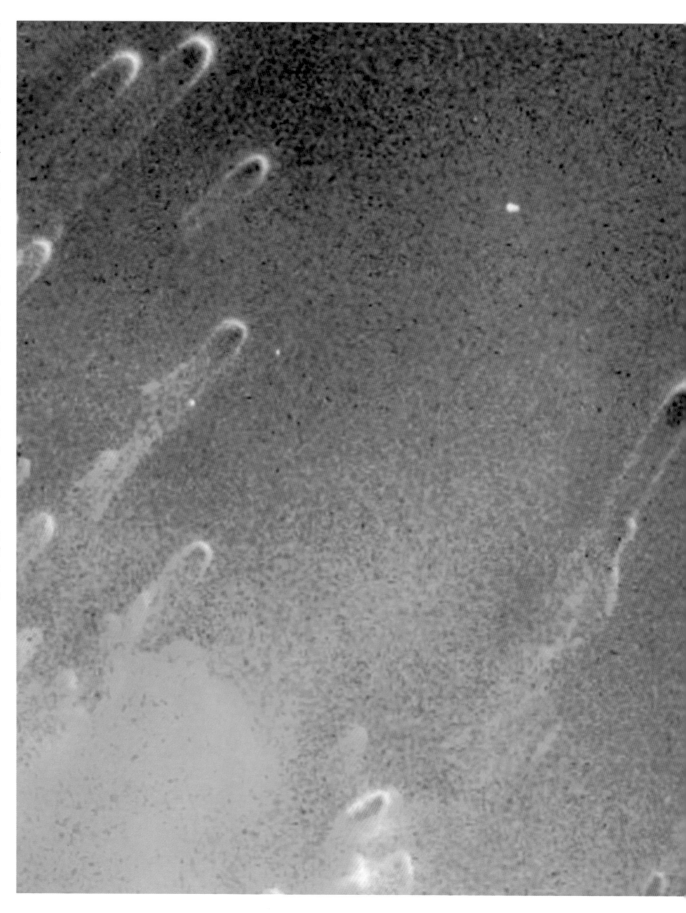

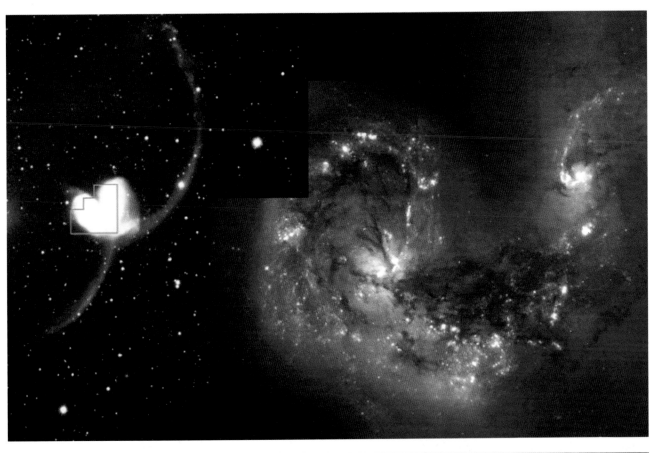

At far left, photographed with an Earth-based telescope, **NGC 4038** and **NGC 4039** are two interacting galaxies 63 million light-years away. Because of the shape of their appendages, they are called the Antennae Galaxies.

At near left, we see their orange-colored nuclear regions joined by filaments of dark dust. Blue globular clusters visible in the spiral arms show signs of star-forming activity spurred by the collision of the two galaxies. Image taken by **Brad Whitmore (STScI)** with the WF/PC-2.

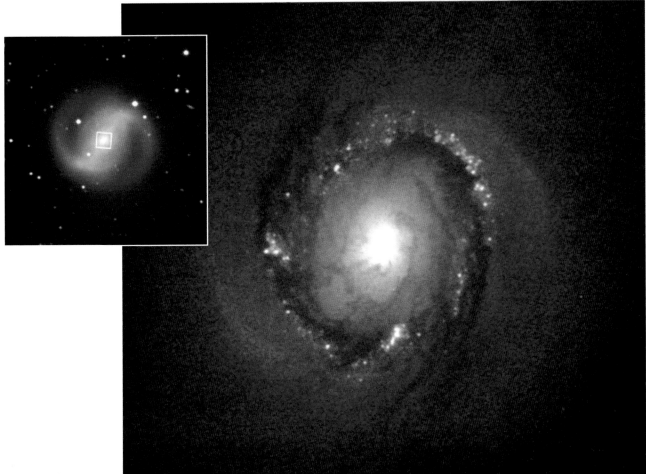

The center of the barred spiral NGC 4314, seen at far left in a picture taken with the 76-cm telescope at McDonald Observatory, presents a very striking aspect in the Hubble WF/PC-2 image at near left. The nucleus is surrounded by a blue and purple ring made up of newborn stars. Outside the ring, two strips of dark dust and a pair of spiral arms emerge, their characteristic blue color indicating youth, though they are older than the inner ring stars. This seems to suggest that the locus of star-forming activity is moving toward the center of the galaxy.

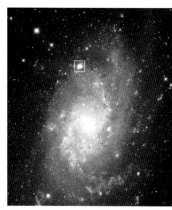

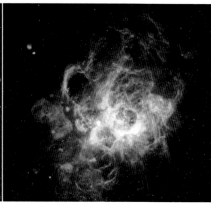

Above right, a Hubble WF/PC-2 image of the nebula NGC 604, a star-forming region in a spiral arm of the galaxy M33 in Triangulum (see page 87). The nebula is so enormous (1,500 light-years in diameter) that it can be seen even in ground-based tele-scope images (above left, boxed object). More than 200 very hot stars lie at the heart of NGC 604. They are 15 to 60 times larger than the Sun, and they heat and illuminate the gases of the cloud, highlight-ing its three-dimension-al form like a lantern in a cave. The study of the physical structure of such giant nebulas can help astronomers understand how star clusters may influence stellar evolution in the arms of spiral galaxies.

Below, M2-9 is a splen-did bipolar, or "butter-fly," planetary nebula. The central star is in fact a very tight binary system; the gravity of one star tugs gas from the surface of the other and propels it into a thin, dense ring encir-cling both stars. High-speed winds emitted by one of the stars collide with the gaseous disk, which acts like a rocket nozzle, deflecting the winds in a direction per-pendicular to the plane of the disk and thereby forming the two spouts visible in the photo-graph.

■ Image taken by B. Balick, V. Icke and G. Mellema with the WF/PC-2.

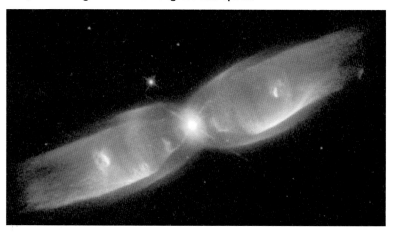

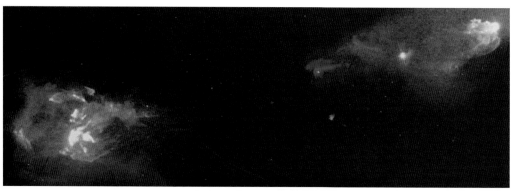

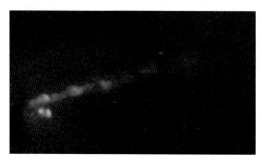

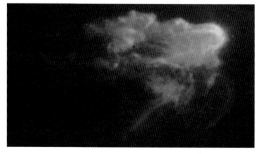

At left, images showing gas jets of Herbi-Haro objects (HH1 and HH2) in Orion, 1,500 light-years away, open new perspectives on stellar formation. In the upper image, we see jets emit-ted by an unseen star at the center that is hid-den behind a dark cloud of dust. Where the jets run into the interstellar gas, they form the lobes visible at the far ends. The lower left image, a detail of the region close to the star, reveals a thin strip formed by puffs of gas emitted at 200 km per second from a vortex of gas and dust surrounding the protostar. To the right, a detail of the arrow-shaped structure attesting to the shock wave produced by the high-velocity jets' colli-sion with the interstel-lar medium.

■ Images taken by J. Hester of the Univer-sity of Arizona with the WF/PC-2.

*After the December 1993 Hubble repair mission, NASA and ESA scientists were understandably anxious as they awaited the first images from the new WF/PC and the FOC corrected by COSTAR. The outcome, however, was exceptional. Not only had the astronauts completed their tasks without error, they had positioned the new devices so precisely that their collimation required only 3 weeks rather than the predicted 9 or 10.*

*In March 1997, a further Hubble servicing mission added two second-generation instruments: the Space Telescope Imaging Spectrograph (STIS)—hundreds of times more powerful than its predecessor—and the Near Infrared Camera and Multi-Object Spectrometer (NICMOS), which extended the Hubble's detection range into the near infrared (for additional NICMOS images, see pages 126-127). Further improvements are*

*planned in the years to come. In 1999, the Faint Object Camera will be replaced by the Advanced Camera for Surveys (ACS), capable of registering images from mid-ultraviolet to near infrared. In 2002, Hubble will get two upgrades: a new Wide Field Camera, the WFC3, able to capture images in near ultraviolet to 200 nanometers, and a Cosmic Origins Spectrograph (COS)—an ultraviolet instrument even more sensi-*

In May 1997, NICMOS photographed the central zone of the Egg Nebula (see page 63), demonstrating the mechanisms by which stars similar to our Sun expel nitrogen and carbon—elements essential to life—into space. In the image at right, the dazzling flash of hydrogen molecules, rendered in red in this false-color enhancement (infrared radiation is invisible to the human eye), shows the region where high-velocity material ejected from the star collides with the slower flow emitted earlier. Below right, a NICMOS image of the central region of the giant molecular cloud OMC1 in Orion (visible at lower left in an image taken with the WF/PC-2) reveals a paroxysm of star formation. Both the stars and the interstellar dust, which scatters and is reheated by stellar radiation, appear to be a yellow-orange color. The brightest object in the image is the young and massive star Becklin-Neigebauer. The bluish strips are channels carved into the cloud by stellar winds from young stars that are probably still enshrouded in dust.

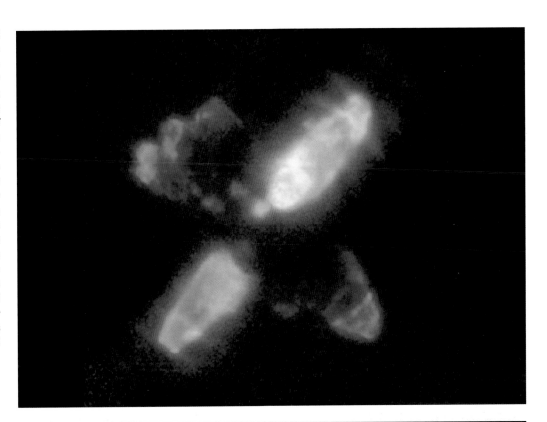

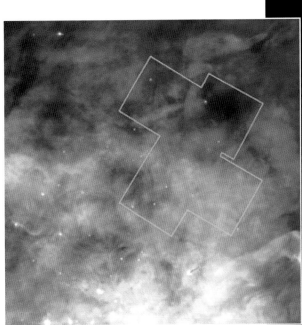

*tive than the STIS. With these five instruments (STIS, NICMOS, ACS, WFC3 and COS) in place, Hubble will carry on until the end of its expected operating life span in 2010.*

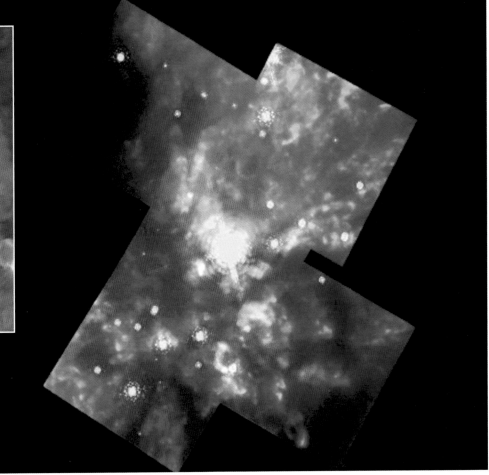

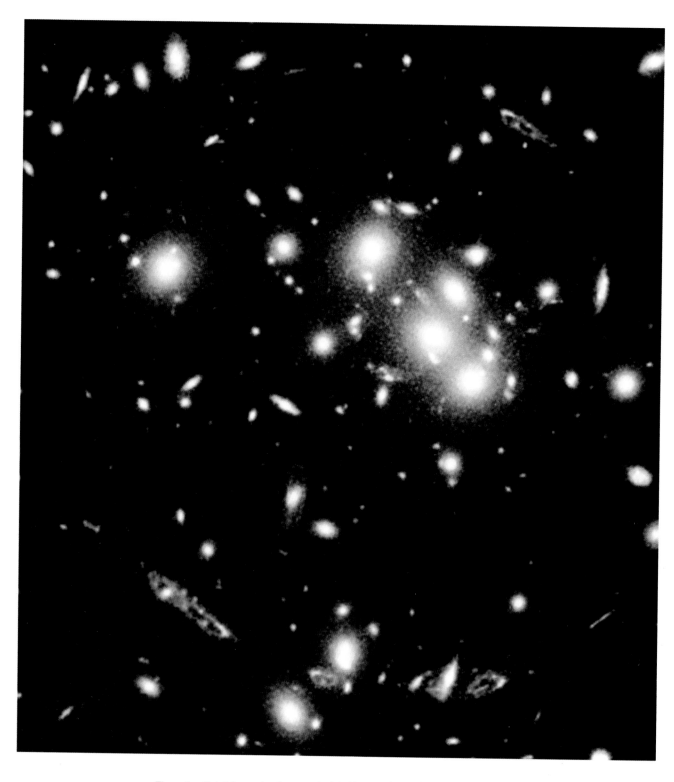

The elongated blue images seen in this photograph are all generated by the same galaxy. They were produced by a gravitational-lens effect stemming from the yellow galactic mass 0024+1654 visible at the center of the picture. Light coming from a galaxy 10 billion light-years away is being deflected in its passage by the nearer galaxy, which lies about half the distance away, producing five distinct images. One is near the center of the photo; the others are at 2, 6, 7 and 8 hours. Gravitational lensing also distorts the galaxy's spiral shape into an arc-like form.
■ Duotone image from WF/PC-2 shots in blue and red light.

The notion that light coming from a celestial object can be deflected by the influence of a gravitational field is a direct result of the theory of general relativity, which posits the equivalence of matter and energy. The first demonstration of this dates to a few years after the theory was published in 1915. British astrophysicist Arthur Eddington succeeded in demonstrating that the positions of a few stars visible near the Sun during the eclipse of 1919 seemed to shift because the light coming from them was being deflected by the Sun's gravitational field.

The first discovery of a gravitational lens—the pair of quasars 0957+561A/B in the constellation Ursa Major—dates from 1980 and was made by English astronomers Dennis Walsh and Robert Carsell and American astronomer Ray Weymann.

Hubble Space Telescope images of quasars—the most distant and luminous objects in the universe—taken with the **WF/PC-2** confirm that these objects reside among a remarkable variety of elliptical and spiral galaxies, most of which are undergoing processes of interaction and collision, leading to the formation of supermassive black holes (see pages 110-117). In many cases, quasars lie exactly halfway between two giant spiral galaxies that are swallowing each other. Many quasars, however, lie at the center of apparently solitary galaxies that seem unperturbed by their presence. Perhaps in these cases, the interacting companion is too close to the galaxy's nucleus to be seen. Or perhaps other mechanisms are at work to spark the fuse that propels the birth of a quasar. If so, it would indicate that on the astronomical time scale, these are short-lived phenomena.

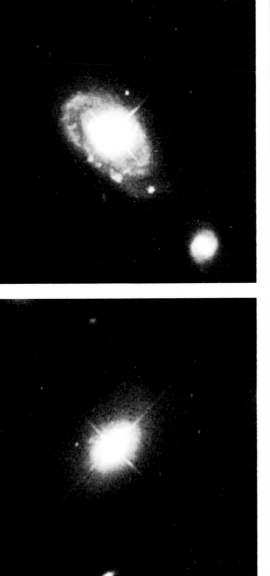

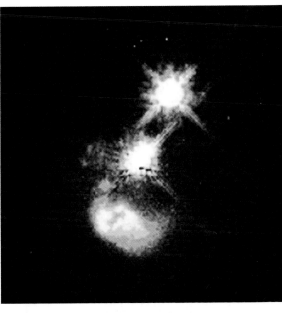

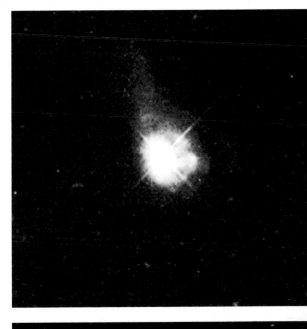

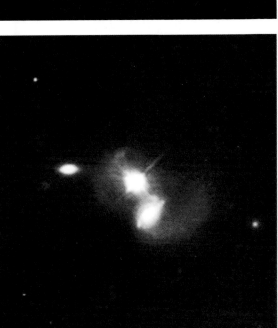

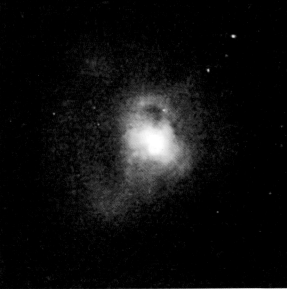

*In 1929, Edwin Hubble discovered that the universe was expanding and that all galaxies move away from us at a speed which increases with distance. He formulated a law, known as Hubble's law, which established a linear relationship between the velocity of recession and the distance, known as the Hubble Constant ($H_0$). It is enough to know the recessional velocity of a celestial object and the value of $H_0$ to work out the distance.*

*The speed is determined by the relationship between the redshift, known as z, and the speed of light. A luminous source moving away appears to turn red; that is, the wavelength of its light increases with its speed. This redshift is measurable by spectrographs. The value of $H_0$, on the other hand, is still uncertain (see page 135).*

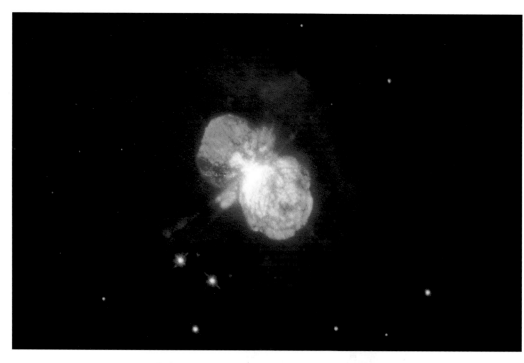

Left, the nebula surrounding the star Eta Carinae (just visible, center, where the telescope's diffraction points cross). Eta Carinae is one of the brightest and most massive stars: 150 times larger than the Sun and 4 million times brighter. The visible gas was ejected during an explosion undergone by the star in the 19th century (see page 58). The outermost reddish halo is made up of nitrogen and other elements formed in the stellar nucleus and is expanding at a rate of 900 km per second. The two white nebular lobes nearest the star are made up of dust that reflects the light of Eta Carinae. The lobe, lower right, is expanding in our direction, while the other one is expanding in the opposite direction. Previous models of this object showed a dense disk surrounding the star, with the ejected material being expelled from the disk's poles. What can be seen instead, unpredictably, is that the matter is scattered at high speed from within the plane of the hypothetical disk itself. ■ Picture taken by Jeff Hester of Arizona State University with the WF/PC-2.

Right, the star-forming region in the 30 Doradus Nebula in the Large Magellanic Cloud. The larger picture is a mosaic created with the WF/PC-2 in wide-field mode; the smaller one is an image of an area within the same field in planetary mode, at maximum resolution. The nebula is an H II region, formed by hydrogen ionized by the intense ultraviolet radiation coming from hot stars. In this case, there are more than 3,000 stars, among the brightest and most massive known, and they form a cluster, known as R 136, clearly visible in the smaller picture. The analysis of this picture of R 136 shows that the formation of stars of a mass much greater than that of the Sun is just as rapid as that of much less massive stars.

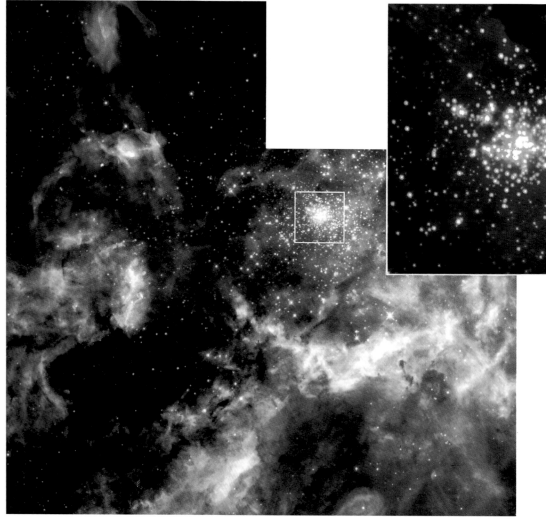

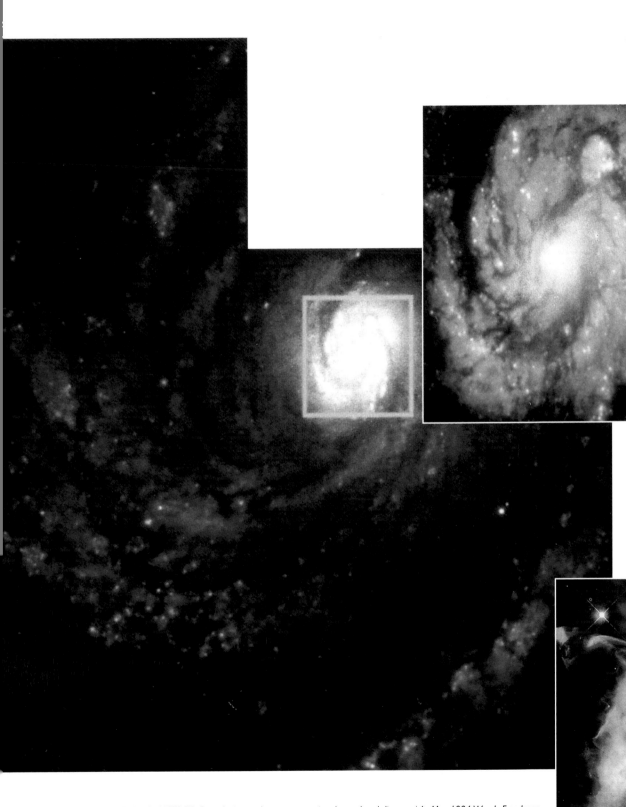

Left, the galaxy **M100** photographed with the **WF/PC-2.** In the wide-field view, various individual stars can be seen, as well as long filaments of dust in the outer spiral arms, while the close-up view taken with the **PC** resolves the complex structure of the galactic nucleus.

■ Composite of three images photographed in blue, green and red light. Below, a region of the Great Nebula in Orion, one of the closest star-forming regions. The various colors correspond to different gases ionized by the ultraviolet radiation of bright stars: red corresponds to nitrogen, green to hydrogen, blue to oxygen. Some of the stars visible are surrounded by disks of dust and gas, which are probably protoplanetary (see page 49).

■ Shot taken with the **WF/PC-2** by **C.R. O'Dell** of Rice University.

The resolution attainable by the WF/PC-2 for galaxies such as M100 has made it possible to address one of the key astronomical objectives for which the Hubble was designed: the observation of Cepheid variables (page 47) in distant galaxies. Such observations will help us to know with precision the distance to the galaxies and from this, according to Hubble's law, to calculate the value of $H_0$. (The latter is simply the relationship between recessional speed and distance.) In May 1994, Wendy Freedman of the Carnegie Observatories studied 20 Cepheids in M100 and was able to measure their distances with precision and, thus, to calculate the value of $H_0$ as 25 km per second per million light-years. This result seems to demonstrate, among other things, that the universe is not as old as had been thought, since it yields an age for the universe of only about 8 billion years.

# The giant telescopes.

**Optical telescopes with a diameter of 3 m or more that are currently operational.**

All of the largest optical telescopes operate with mirrors, employing either a Cassegrain system of optics or a variant of the Cassegrain system known as Ritchey-Chrétien. A good example is the Hubble Space Telescope (page 118): Light enters the telescope tube at one end and is reflected off the primary mirror to a secondary mirror that then reflects the light back through a hole in the center of the primary mirror to converge at the telescope's focal point, the location for its detection devices—cameras, spectroscopes and CCDs.

The focal ratio of a telescope is the mathematical relationship between the focal length of the primary mirror (the distance from the mirror's surface to its focal point, where the light rays converge) and its diameter.

Most large telescopes have more than one focal ratio because besides the Cassegrain focus, they also have a "prime" focus at approximately the position of the secondary mirror, and they may have one or more longer focuses, obtained by interposing additional tertiary mirrors.

| Telescope<br>Observatory<br>Location | Diameter<br>Optical system<br>f-number | Latitude<br>Longitude<br>Altitude | Year | Comments |
|---|---|---|---|---|
| **Keck I Telescope**<br>W.M. Keck Observatory<br>Mauna Kea, Hawaii, USA | 9.82 m<br>Ritchey-Crétien<br>f/1.75-15-25 | 19°49' N<br>155°28' W<br>4150 m | 1991 | Active optics.<br>Hexagonal<br>segments. |
| **Bol'shoi Teleskop Azimutal'nyi**<br>Special Astrophysical Observatory<br>Mount Pastukhov, Russia | 6 m<br>Ritchey-Crétien<br>f/4-30 | 43°39' N<br>41°26' E<br>2100 m | 1975 | Primary mirror<br>replaced in 1984. |
| **George Ellery Hale Telescope**<br>Palomar Observatory<br>Mount Palomar, California, USA | 5.08 m<br>Cassegrain<br>f/3.3-16-30 | 33°21' N<br>116°52' W<br>1706 m | 1948 | |
| **Multiple Mirror Telescope**<br>MMT Observatory<br>Mount Hopkins, Arizona, USA | 4.5 m<br><br>f/2.7-9 | 31°41' N<br>110°53' W<br>2608 m | 1979 | Six individual<br>mirrors. |
| **William Herschel Telescope**<br>Roque de Los Muchachos<br>Observatory<br>La Palma, Canary Is., Spain | 4.2 m<br><br>f/2.5-11 | 28°46' N<br>17°53' W<br>2332 m | 1987 | Property of<br>the Royal<br>Greenwich Obs. |
| **4-m Telescope**<br>Cerro Tololo Inter-American<br>Observatory<br>Cerro Tololo, Chile | 4 m<br>Ritchey-Crétien<br>f/2.8-8 | 30°10' S<br>70°49' W<br>2215 m | 1976 | |
| **Anglo-Australian Telescope**<br>Anglo-Australian Observatory<br>Siding Spring, Australia | 3.89 m<br>Ritchey-Crétien<br>f/3.3-8-15-36 | 31°17' S<br>49°04' E<br>1149 m | 1975 | |
| **Nicholas Mayall Reflector**<br>Kitt Peak National Observatory<br>Kitt Peak, Arizona, USA | 3.81 m<br>Ritchey-Crétien<br>f/2.7-8-15.7-190 | 31°58' N<br>111°36' W<br>2120 m | 1973 | |
| **United Kingdom Infrared Telescope**<br>Joint Astronomy Centre<br>Mauna Kea, Hawaii, USA | 3.8 m<br>Cassegrain<br>f/2.5-36 | 19°50' N<br>155°28' W<br>4194 m | 1978 | Infrared only. |
| **Canada-France-Hawaii Telescope**<br>Canada-France-Hawaii Telescope Corp.<br>Mauna Kea, Hawaii, USA | 3.58 m<br>Cassegrain<br>f/3.8-8-20-35 | 19°49' N<br>155°28' W<br>4200 m | 1979 | |
| **Telescopio Nazionale Galileo**<br>Roque de Los Muchachos<br>Observatory<br>La Palma, Canary Islands, Spain | 3.58 m<br>Ritchey-Crétien<br>f/2.2-11 | 28°46' N<br>17°53' W<br>2332 m | 1996 | Active and<br>adaptive optics. |
| **ESO 3.6-m Telescope**<br>European Southern Observatory<br>La Silla, Chile | 3.57 m<br>Ritchey-Crétien<br>f/3-8.1-32-35 | 29°16' S<br>70°44' W<br>2387 m | 1977 | |
| **3.5-m Telescope**<br>Observatorio de Calar Alto<br>Calar Alto, Spain | 3.5 m<br>Ritchey-Crétien<br>f/3.5-3.9-10-35 | 37°13' N<br>2°32' W<br>2168 m | 1984 | Property of the<br>German-Spanish<br>Astronomical Center. |

| Telescope / Observatory / Location | Diameter / Optical system / f-number | Latitude / Longitude / Altitude | Year | Comments |
|---|---|---|---|---|
| **New Technology Telescope** / European Southern Observatory / La Silla, Chile | 3.5 m / Ritchey-Crétien / f/2.2-11 | 29°16' S / 70°44' W / 2353 m | 1989 | Active optics. |
| **Starfire Optical Range Telescope** / Kirtland Air Force Base / New Mexico, USA | 3.5 m / / f/1.5 | 35°02' N / 106°37' W / 1950 m | 1994 | Adaptive optics. |
| **Astrophysical Research Consortium Telescope** / Apache Point, New Mexico, USA | 3.5 m / / f/1.75 | 32°47' N / 105°49' W / 2800 m | 1994 | |
| **Wisconsin-Indiana Yale-NOAO Telescope** / WIYN Observatory / Kitt Peak, Arizona, USA | 3.5 m / Ritchey-Crétien / f/1.75-6.3 | 31°57' N / 111°36' W / 2089 m | 1994 | Active optics. |
| **Donald Shane Telescope** / Lick Observatory / Mount Hamilton, California, USA | 3.05 m / Ritchey-Crétien / f/5-17-36 | 37°21' N / 121°38' W / 1290 m | 1959 | |
| **NASA Infrared Telescope Facility** / Mauna Kea Observatory / Mauna Kea, Hawaii, USA | 3 m / Cassegrain / f/2.5-35-120 | 19°50' N / 155°28' W / 4208 m | 1979 | Infrared only. |

**Schmidt telescopes with a diameter of 1 m or more currently operational**

| Telescope / Observatory / Location | Diameter / Optical system / f-number | Latitude / Longitude / Altitude | Year | Comments |
|---|---|---|---|---|
| **2-m Telescope** / Karl Schwarschild Observatorium / Tautenberg, Germany | 1.34 m / / f/3.0 | 50°59' N / 11°43' E / 331 m | 1960 | Also Cassegrain. |
| **Oschin 48-in Telescope** / Palomar Observatory / Mount Palomar, California, USA | 1.22 m / / f/2.47 | 33°21' N / 116°51' W / 1706 m | 1948 | |
| **United Kingdom Schmidt Telescope Unit** / Royal Observatory, Edinburgh / Mount Siding Spring, Australia | 1.22 m / / f/2.5 | 31°16' S / 149°04' E / 1145 m | 1973 | |
| **Kiso Schmidt Telescope** / Kiso Observatory / Kiso, Japan | 1.05 m / / f/3.1 | 35°48' N / 137°38' E / 1130 m | 1975 | Optional secondary f/22.6. |
| **3TA-10 Schmidt Telescope** / Astrophysical Observatory of Byurakan / Mount Aragatz, Armenia | 1.0 m / / f/2.13 | 40°20' N / 44°30' E / 1450 m | 1961 | |
| **Kvistaberg Schmidt Telescope** / Observatory of the University of Uppsala / Kvistaberg, Sweden | 1.0 m / / f/3.0 | 59°30' N / 17°36' E / 33 m | 1963 | |

| Telescope / Observatory / Location | Diameter / Optical system / f-number | Latitude / Longitude / Altitude | Year | Comments |
|---|---|---|---|---|
| **ESO 1-m Schmidt Telescope** / European Southern Observatory / La Silla, Chile | 1.0 m / / f/3.06 | 29°15' S / 70°44' W / 2318 m | 1972 | |
| **Telescopio Schmidt da 1 m del Venezuela** / Centro "F.J. Duarte" / Llano del Hato, Merida, Venezuela | 1.0 m / / f/3.0 | 8°47' N / 70°52' W / 3610 m | 1978 | |

**Optical telescopes with diameter over 3 m currently under construction**

| Telescope / Observatory / Location | Diameter / Optical system / f-number | Latitude / Longitude / Altitude | Year | Comments |
|---|---|---|---|---|
| **Very Large Telescope** / European Southern Observatory / Cerro Paranal, Chile | 16 m (equivalent) / Ritchey-Crétien / f/13.5-15 (each mirror) | 24°51' S / 70°27' W / 2640 m | 2000 | 4 mirrors 8.2 m each, adaptive optics. |
| **Large Binocular Telescope** / Columbus Project / Mount Graham, Arizona, USA | 11.8 m (equivalent) / Cassegrain / f/1.14-5.4-15 (each mirror) | 32°42' N / 109°51' W / 3170 m | 1997 | 2 mirrors 8.4 m each. |
| **Hobby-Eberly Telescope** / McDonald Observatory / Mount Locke, Texas, USA | 11 m / / f/1.3 | 30°40' N / 104°01' W / 2075 m | 1996 | Hexagonal segments, spectroscope only. |
| **Keck II Telescope** / W.M. Keck Observatory / Mauna Kea, Hawaii, USA | 9.82 m / Ritchey-Crétien / f/1.75-15-25 | 19°49' N / 155°28' W / 4150 m | 1996 | Similar to Keck I |
| **Subaru Telescope** / National Astronomical Observatory (Japan) / Mauna Kea, Hawaii, USA | 8.3 m / Ritchey-Crétien / f/1.8-12.5-35 | 19°49' N / 155°28' W / 4215 m | 2000 | Active optics. |
| **Gemini Telescope (North)** / Joint Astronomy Center / Mauna Kea, Hawaii, USA | 8.1 m / / f/1.8-16-19.6 | 19°49' N / 155°28' W / 4100 m | 1999 | Optimized for infrared. |
| **Gemini Telescope (South)** / Cerro Tololo Inter-American Observatory / Cerro Pachon, Chile | 8.1 m / / f/1.8-6-16-19.6 | 30°21' S / 70°49' W / 2725 m | 2001 | Jointly owned by the USA, Argentina, UK, Chile, Brazil and Canada. |
| **New Multiple Mirror Telescope** / Mount Hopkins, Arizona, USA | 6.5 m / / f/1.25-5.4-9-15 | 31°41' N / 110°53' W / 2608 m | 1996 | |
| **Magellan Project** / Las Campanas Observatory / Las Campanas, Chile | 6.5 m / Cassegrain and Gregory / f/1.25-11-15 | 29°00' S / 70°42' W / 2300 m | 1998 | Property of the Carnegie Institution, Washington. |

**Glossary of technical terms that are not explained in the text.**

**Bar** Elongated structure that is found in some spiral galaxies instead of a nuclear bulge, giving them a particular shape known as "barred."

**Composite image** Photographic technique that involves printing two or more negative images of the same subject and then superimposing the prints in order to lessen their graininess and to emphasize the contrast and visibility of details.

**False-color image** Computer-enhancement technique employing black-and-white photographs, in which different colors are arbitrarily assigned to the various gray tones in order to improve the quality of pictures and to show up specific details more clearly. When the process involves only one monochromatic plate, it is known as *pseudocolor*. When it is carried out with several plates taken in different spectral wavelengths, it is quite appropriately known as a *false-color* enhancement.

**Flyby** Astronautical technique in which a space probe is made to fly as close as possible to its target. With it, exploration is limited to just those few hours in which the greatest degree of proximity is achieved. More preferable but much more expensive is the so-called *rendezvous*, in which the probe travels for a long time alongside the body it is to explore.

**Image intensifier** Electronic device fitted onto telescopes which amplifies starlight by a few dozen times, enabling a drastic reduction in the exposure times necessary to capture images of celestial objects on film.

**Magnitude** Measure of the brightness of a celestial object. *Apparent magnitude* is the observed brightness (without taking into account the distance of the object). *Absolute magnitude* is the brightness an object would show if placed at a standard distance of 10 parsecs (32.6 light-years). The latter expresses the true, intrinsic luminosity of the object. Magnitude ranges from negative or very low values for the brightest objects up to very high values for the faintest objects. The brightest star visible to the naked eye in the night sky, Sirius, has an apparent magnitude of -1.46. The faintest stars visible to the naked eye have an apparent magnitude of 6. The faintest ones visible with the most sophisticated telescopes have an apparent magnitude of 29.

**NGC** stands for New General Catalog, a catalog of 7,840 nebular objects (galaxies, nebulas and clusters), published by astronomer Johan Ludwig Emil Dreyer in 1888 and integrated by him in 1895 and 1908 into the Index Catalog (IC).

**Orbiter** Space vehicle that is put into a stable orbit around a planet or other celestial object. In this way, exploration can be more complete and detailed, in comparison with the *flyby* and *rendezvous* techniques.

**Planetesimals** Chunks of rock and ice a few kilometers in diameter which, by aggregating, led, 4.6 billion years ago, to the formation of the planets in our solar system.

**Protostar** Cloud of interstellar gas which, after contracting, reaches the necessary pressure and temperature to trigger the thermonuclear reactions that lead to the birth of a star.

**Radio galaxy** Galaxy that, besides emitting visible light, also releases radio waves, resulting from the high-speed emission of electrons in very intense magnetic fields. This emission is known as synchrotron radiation, because of the resemblance with the emissions produced in particle accelerators, or synchrotrons.

**Seeing** Term indicating the conditions of astronomical observation, essentially by measuring the maximum angular resolution that can be achieved. On the ground, seeing is heavily affected by atmospheric turbulence; in space, it depends exclusively on the diameter and effective accuracy of the telescope's optics. For a terrestrial telescope, seeing with a resolution of 1.0 arc-second (equivalent to seeing a crater 1.9 km in diameter on the Moon) is good; seeing of 0.5 arc-second is very good; and finally, seeing of 0.2 to 0.3 arc-second is excellent. The Hubble Space Telescope achieves an average resolution of 0.05 arc-second, which is the same as picking out a crater on the Moon less than 100 m in diameter.

**Tidal force** A disturbance that occurs in celestial objects that are close to one another, resulting from the difference of intensity in the gravitational forces acting on the nearer and more distant parts of the objects. Tidal forces cause fluids, such as the Earth's oceans or the gases in a galaxy, to undergo considerable deformations of shape.

**Key to acronyms used here and in the text:**
**AATB: Anglo-Australian Telescope Board**
**Caltech: California Institute of Technology**
**ESA: European Space Agency**
**ESO: European Southern Observatory**
**JPL: Jet Propulsion Laboratory**
**NASA: National Aeronautics and Space Administration**
**NOAO: National Optical Astronomy Observatories**
**STScI: Space Telescope Science Institute**
**USGS: United States Geological Survey**

AATB: 17 (top)
R. Albrecht, ESA/ESO Space Telescope European Coordinating Facility/NASA: 34
American Science and Engineering/Harvard College Observatory: 40 (bottom)
Bruce Balick (University of Washington)/Vincent Icke (University of Leida), Garrelt Mellema (University di Stockholm)/NASA: 132 (center)
Reta Beebe, Amy Simon (New Mexico State Univ.)/NASA: 124 (bottom)
G. Fritz Benedict, Andrew Howell, Inger Jorgensen, David Chapell (University of Texas)/ Jeffrey Kenney (Yale University)/ Beverly J. Smith (CASA, University of Colorado)/NASA: 129 (bottom)
Matt Bobrowsky (Orbital Sciences Corp.), Sahu (STScI), M. Parthasarathy (Indian Institute of Astrophysics), Pedro Garcia-Lario (ISO Science Operation Center)/NASA/ESA: 62 (bottom)
H. Bond (STScI)/NASA: 63 (top right)
Kirk D. Borne, STScI/NASA: 109 (bottom left)
Claudia Burg/VATT: 94 (top)
Cristopher Burrows, ESA/STScI/NASA: 67 (inset)
David F. Buscher, Institute of Astronomy, Cambridge University: 42 (bottom)
Caltech/University of California: 16 (right)
W.N. Colley, E. Turner (Princeton University)/ J.A. Tyson (Bell Labs)/NASA: 130
Conservatorio d'arti e mestieri, Parigi: 9 (bottom)
Emmanuel Davoust, Observatoire Midi-Pyrénées: 114
Rebecca Elson, Richard Sword (Cambridge, UK)/ James Westphal (Caltech)/NASA: 78 (top left)
ESA: 17 (bottom right), 36, 119
ESA/NASA: 118, 121 (top),
ESO: 13, 14 (bottom), 15 (center), 16 (top), 35, 54, 56, 58, 65 (top), 66, 77 (top), 79, 92, 95, 99 (bottom), 110, 111 (center)
Fred Espenak/Science Photo Library/Grazia Neri: 39 (top)
H. Ford, Z. Tsvetanov, A. Davidsen, G. Kriss (Johns Hopkins Univ.); R. Harms, L. Dressel, A.K. Kochhar (Applied Res. Corp.); R. Bohlin, G. Hartig, STScI; B. Margon, Univ. of Washington/NASA: 113
Robert Q. Fugate/Starfire Optical Range-USAF Phillips Laboratory: 14 (left)

Akira Fujii: 82, 84
R. Gilmozzi (STScI/ESA)/ Shan Ewald (JPL)/NASA: 78 (bottom left)
Eraldo Guidolin: 37 (right)
Hale Observatories, Caltech: 61 (inset), 65 (bottom), 68, 72-73, 93 (bottom), 99 (top)
Tony and Daphne Hallas: 51, 52-53, 86
J.P. Harrington, K.J. Borkowski (University of Maryland)/NASA: 63 (bottom left)
Harvard-Smithsonian Center for Astrophysics: 44 (right), 114 (inset)
J. Hester, Arizona State University/NASA: 132 (bottom), 134 (top)
Jeff Hester, Paul Scowen (Arizona State University)/NASA: 57 (bottom)
John Holtzman, Lowell Observatory/WFPC Team/NASA/ESA: 127 (top)
Hubble Space Telescope Comet Team/NASA: 127 (bottom)
Walter Jaffe, Leyden Observatory/Holland Ford, Johns Hopkins University/NASA/ESA: 113 (bottom right)
Robert Jedrzejewski/NASA/ESA: 78 (bottom)
JPL/University of Arizona: 12
Erich Karkoschka (University of Arizona)/NASA: 126 (top), 127 (top)
L'Astronomia: 11
Severino Lodi: 37 (left)
Duccio Macchetto/ESA/STScI/FOC Team/NASA/ESA: 115 (right)
David Malin/AATB: 49 (top), 50 (bottom), 52 (inset), 55, 57 (top), 60, 61, 64, 67, 70, 74 (left), 88, 89, 90-91, 96, 97 (top), 112 (right)
David Malin/AATB/Royal Observatory Edinburgh: 50 (top), 59, 69, 98, 100, 102-103
David Malin/Instituto de Astrofisica de Canarias: 93 (top)
J. Mazzarella: 111 (right)
Robert McClure, National Research Council of Canada: 14 (center)
Museo di Storia della Scienza, Florence: 8 (left)
NASA: 24, 29 (bottom), 120, 121 (bottom), 123 (top)
NASA Ames Research Center: 21
NASA Goddard Space Flight Center: 17 (bottom right), 14 (bottom), 82 (inset), 105 (bottom), 117 (bottom), 125
NASA/JPL: 17 (left), 19, 20 (left), 22, 25, 26 (bottom, right and left), 27, 28, 30, 31 (top and bottom right), 32, 33, 49 (bottom), 134 (bottom)
NASA/JPL/Science Photo Library/Grazia Neri: 20 (right)
NASA/JPL/USGS: 23
NASA/JPL/USGS/Cornell University: 26 (top)
NASA Lyndon Johnson Space Center: 122, 123 (bottom), 135 (top)
NASA/STScI: 112 (inset)
National Radio Astronomy Observatory/Associated Universities, Inc.: 83

Naval Research Laboratory: 41 (top)
NOAO: 16 (center), 40 (top), 62 (top), 71, 77 (bottom, left), 91, 94 (bottom), 103, 104 (top), 108 (bottom), 109 (top and bottom right), 116, 117 (top)
NOAO/T. Boroson: 107
NOAO/W. Schoening/N. Sharp: 87 (right)
NOAO/N.A. Sharp: 97 (bottom), 108 (top)
C.R. O'Dell/Rice University/NASA: 49 (left), 135 (inset)
Robert O'Dell, Kerry Handron (Rice University)/ NASA: 128
Tom Prentiss: 9 (top)
Roger H. Ressmeyer-Starlight/MPA: 15 (left)
Michael Rich, Kenneth Mighell, James D. Neill (Columbia University), Wendy Freedman (Carnegie Observatories)/NASA: 78 (top right)
Raghvendra Sahai, John Trauger (JPL)/WFPC2 science team/ NASA: 63 (top left and bottom right)
Don Savage (NASA)/Tammy Jones (Goddard Space Flight Center)/Ray Villard (STScI)/John Bahcall (Institute for Advanced Study)/ Barbara Kennedy (Pennsylvania State Univ.)/NASA: 131
Rudolph E. Schild: 111 (top)
Peter H. Smith, University of Arizona Lunar and Planetary Laboratory/NASA: 31 (bottom left)
Thomas Soifer/Caltech/University of California: 115 (bottom left)
Alan Stern (Southwest Research Institute)/Marc Buie (Lowell Observatory)/NASA/ESA: 34 (bottom)
STScI: 110
Rodger Thompson, Marcia Rieke, Glenn Schneider, Dean Hines (University of Arizona)/Raghvendra Sahai (JPL)/NICMOS Instrument Definition Team/ NASA: 133 (top)
Robert Thomson/NASA/ESA: 115 (top)
USGS: 24
U.S. Naval Observatory: 76
Gabriele Vanin: 39 (bottom), 48, 75 (top, bottom), 80 (left), 81 (top)
Gabriele Vanin/Associazione Astronomica Feltrina *Rheticus*: 42 (top), 43, 44 (left), 45, 46, 47, 73 (inset), 74 (right), 75 (center)
Ray Villard (STScI)/NASA: 124 (top)
Harold Weaver (Johns Hopkins University)/NASA: 126 (center)
H.A. Weaver and T.E. Smith, STScI/NASA: 126 (bottom)
James Westphal, Caltech/WFPC Team/NASA/ESA: 125
Brad Whitmore, STScI/NASA/ESA: 129 (top)
Robert Williams/Hubble Deep Field Team (STScI)/ NASA: 104 (bottom)
Hui Yang (University of Illinois)/NASA: 132 (top)

The publisher apologizes for any errors or omissions from the list of photo sources.

Anderson, Charlene M. *Europa: a world of superlatives,* "The Planetary Report" vol. 12, no. 1 (January-February 1992), p. 12.

Ashbrook, Joseph. *Early photography and the great comet of 1882,* "Sky & Telescope" vol. 22, no. 6 (December 1961), p. 331.

Ashbrook, Joseph. *An episode in early astro-photography,* "Sky & Telescope" vol. 57 no. 2 (February 1979), p. 141.

Audouze, Jean; Israîl, Guy (ed.). *The Cambridge atlas of astronomy,* 2nd ed., Cambridge (UK), Cambridge University Press, 1988.

Bahcall, John N.; Spitzer, Lyman Jr. *Lo Space Telescope,* "Le Scienze" no. 169 (September 1982), p. 10.

Balsiger, Hans; Fechtig, Hugo; Geiss, Johannes. *Uno sguardo ravvicinato alla cometa di Halley,* "Le Scienze" no. 243 (November 1988), p. 74.

Barnes, Joshua; Hernquist, Lars; Schweizer, François. *Galassie in collisione,* "Le Scienze" no. 278 (October 1991), p. 28.

Beatty, Kelly J.; Chaikin, Andrew (ed.). *The new solar system,* 3rd ed., Cambridge (USA) and Cambridge (UK), Sky Publishing Corporation and Cambridge University Press, 1990.

Beatty, Kelly J. *Working Magellan's magic,* "Sky & Telescope" vol. 86, no. 2 (August 1993), p. 16.

Beatty, Kelly J.; Levy, David. *Awaiting the crash I & II,* "Sky & Telescope" vol. 87, no. 1 and vol. 88, no. 1 (January and July 1994), pp. 40 and 18.

Binzel, Richard P. *Plutone,* "Le Scienze" no. 264 (August 1990), p. 14.

Blitz, Leo, et al. *The centre of the Milky Way,* "Nature" vol. 361, no. 6411 (4 February 1993), p. 417.

Böhm, Conrad A. *Il leviatano di Parsonstown,* "L'Astronomia" no. 34, (June 1984), p. 42.

Böhm, Conrad A. *Le chiavi del cosmo,* Padua, Muzzio, 1989.

Braccesi, Alessandro; Caprara, Giovanni; Hack, Margherita. *Alla scoperta del sistema solare,* Milan, Mondadori, 1993.

Bruning, David. *Hubble better than new,* "Astronomy" vol. 22, no. 4 (April 1994), p. 44.

Burbidge, Geoffrey; Hewitt, Adelaide. *A catalog of quasars near and far,* "Sky & Telescope" vol. 88, no. 6 (December 1994), p. 32.

Burnham, Robert Jr. *Burnham's celestial handbook,* New York, Dover, 1978.

Burns, Jack O. *Strutture grandissime nell'universo,* "Le Scienze" no. 217 (September 1986), p. 18.

Chaffee, Frederic H. Jr. *La scoperta di una lente gravitazionale,* "Le Scienze" no. 149 (January 1981), p. 36.

Chevalier, Roger A. *Supernova 1987 A at five years of age,* "Nature" vol. 355, no. 6362 (20 February 1992), p. 691.

Chin, Y.-N.; Huang, Y.-L. *Identification of the guest star of AD 185 as a comet rather than a supernova,* "Nature" vol. 371, no. 6496 (29 September 1994), p. 398.

Clark, David H.; Stephenson, F. Richard. *The historical supernovae,* Oxford, Pergamon Press, 1977.

Courvoisier, Thierry J.-L.; Robson, E. Iano. *Il Quasar 3C 273,* "Le Scienze" no. 276 (August 1991), p. 20.

Davidson, Greg. *How we'll fix Hubble,* "Astronomy" vol. 21, no. 2, (February 1993), p. 42.

De Vaucouleurs, Gerard. *I bracci di M 31,* "L'Astronomia" no. 80, (August-September 1988), p. 18.

Elmegreen, Debra Meloy; Elmegreen, Bruce. *What puts the spiral in spiral galaxies?* "Astronomy" vol. 21, no. 9 (September 1993), p. 34.

Falomo, Renato. *Gli oggetti BL Lac e le loro galassie ospiti,* "L'Astronomia" no. 121 (May 1992), p. 2.

Farinella, Paolo. *Pianetini visti da vicino,* "L'Astronomia" no. 150 (January 1995), p. 14.

Favero, Giancarlo. *L'evoluzione del sistema solare,* Rome, Curcio, 1986.

Ferris, Timothy. *Galassie,* Milan, Fabbri, 1981.

Fienberg, Richard Tresch. *HST: astronomy's discovery machine,* "Sky & Telescope" vol. 79 no. 4 (April 1990), p. 366.

Fienberg, Richard Tresch. *Hubble's image restored,* "Sky & Telescope" vol. 87, no. 4 (April 1994), p. 20.

Finkbeiner, Anno. *Active galactic nuclei: sorting out the mess,* "Sky & Telescope" vol. 84, no. 2 (August 1992), p. 138.

Fischer, Daniel. *A telescope for tomorrow,* "Sky & Telescope" vol. 78, no. 3 (September 1989), p. 248.

Francis, Peter. *I pianeti: dieci anni di scoperte,* Torino, Boringhieri, 1985.

Freedman, Wendy L. *Velocità di espansione e dimensioni dell'universo,* "Le Scienze" no. 293 (January 1993), p. 16.

Freedman, Wendy L., et al. *Distance to the Virgo cluster galaxy M 100 from Hubble Space Telescope observations of cepheids,* "Nature" vol. 371, no. 6500 (27 October 1994), p. 757.

Fugate, Robert Q., et al. *Measurement of atmospheric wavefront distortion using scattered light from a laser guide-star,* "Nature" vol. 353, no. 6340 (12 September 1991), p. 144.

Fugate, Robert Q.; Wild, Walter J. *Untwinkling the stars I & II,* "Sky & Telescope" vol. 87, nos. 5 and 6 (May and June 1994), pp. 24 and 20.

Gilardi, Ando. *Storia sociale della fotografia,* Milan, Feltrinelli, 1976.

Gingerich, Oweno. *The first photograph of a nebula,* "Sky & Telescope" vol. 60, no. 4 (October 1980), p. 364.

Godoli, Giovanni. *Il Sole,* Torino, Einaudi, 1982.

Goldstein, Alano. *Observing the Andromeda galaxy,* "Astronomy" vol. 19, no. 11 (November 1991), p. 76.

Goldwurm, A., et al. *Evidence against a massive black hole at the galactic center,* "Nature" vol. 371, no. 6498 (13 October 1994), p. 589.

Greco, Vincenzo; Molesini, Giuseppe; Quercioli, Franco. *Optical tests of Galileo's lenses,* "Nature" vol. 358, no. 6382 (9 July 1992), p. 101.

Guaita, Cesare. *Meraviglie dallo Space Telescope,* "L'Astronomia" no. 127 (December 1992), p. 4.

Hallas, Tony; Mount, Daphne. *Enhanced-color astrophotography,* "Sky & Telescope" vol. 78, no. 2 (August 1989), p. 216.

Hardy, John W. *Ottica adattativa,* "Le Scienze" no. 312 (August 1994), p. 46.

Harris, William E. *Globular clusters in distant galaxies,* "Sky & Telescope" vol. 81, no. 2 (February 1991), p. 148.

Hirshfeld, Alan; Sinnott, Roger W; Ochsenbein, François. *Sky catalogue 2000.0* Vol. 1, 2nd ed., Cambridge (USA) and Cambridge (UK), Sky Publishing Corporation and Cambridge University Press, 1991.

Hirshfeld, Alan; Sinnott, Roger W. *Sky catalogue 2000.0* Vol. 2, Cambridge (USA) and Cambridge (UK), Sky Publishing Corporation and Cambridge University Press, 1985.

Hoskin, Michael. *William Herschel e la nascita dell'astronomia moderna,* "Le Scienze" no. 216 (August 1986), p. 84.

Ibata, R.A.; Gilmore, G.; Irwin, M.G. *A dwarf satellite galaxy in Sagittarius,* "Nature" vol. 370, no. 6486 (21 July 1994), p. 194.

Janesick, James; Blouke, Morley. *Sky on a chip: the fabulous CCD,* "Sky & Telescope" vol. 74, no. 3 (September 1987), p. 238.

Johnson, Torrence V.; Soderblom, Laurence A. *Io,* "Le Scienze" no. 186 (February 1984), p. 50.

Keel, William C. *Crashing galaxies, cosmic fireworks,* "Sky & Telescope" vol. 77, no. 1 (January 1989), p. 18.

King, Henry C. *The history of the telescope,* New York, Dover, 1979.

Kristian, Jerome; Blouke, Morley. *Strumenti microelettronici in astronomia,* "Le Scienze" no. 172 (December 1982), p. 20.

Lada, Charles. *Deciphering the mysteries of stellar origins*, "Sky & Telescope" vol. 85, no. 5 (May 1993), p. 18.

Lake, George. *Understanding the Hubble sequence*, "Sky & Telescope" vol. 83, no. 5 (May 1992), p. 515.

Lake, George. *Cosmology of the Local Group*, "Sky & Telescope" vol. 84, no. 6 (December 1992) p. 613.

Lavega, Sanchez A., et al. *The great white spot and disturbances in Saturn's equatorial atmosphere during 1990*, "Nature" vol. 353, no. 6343 (3 October 1991), p. 397.

Levy, David. *Pearls on a string*, "Sky & Telescope" vol. 86, no. 1 (July 1993), p. 38.

Luhmann, Janet G.; Pollack, James B.; Colin, Lawrence. *La missione Pioneer Venus*, "Le Scienze" no. 310 (June 1994), p. 68.

Maffei, Paolo. *I mostri del cielo*, Milan, Mondadori, 1976.

Maffei, Paolo. *La cometa di Halley*, Milan, Mondadori, 1984.

Malin, David F.; Zealey, William J. *Astrophotography with unsharp masking*, "Sky & Telescope" vol. 57, no. 4 (April 1979), p. 355.

Malin, David F. *Colour photography in astronomy*, "Vistas in astronomy" vol. 24 (1980), p. 219.

Malin, David F. *Improved techniques for astrophotography*, "Sky & Telescope" vol. 62, no. 1 (July 1981), p. 4.

Malin, David F. *A view of the universe*, Cambridge (USA) and Cambridge (UK), Sky Publishing Corporation and Cambridge University Press, 1993.

Malin, David F. *Un universo a colori*, "Le Scienze" no. 302 (October 1993), p. 40.

Mallas, John H; Kreimer, Evered. *The Messier album*, Cambridge (USA), Sky Publishing Corporation, 1978.

Maran, Stephen (ed.). *The astronomy and astrophysics encyclopedia*, New York and Cambridge (UK), Van Nostrand and Cambridge University Press, 1992.

Miller, Joseph S. *La struttura delle nebulose a emissione*, "Le Scienze" no. 77 (January 1975), p. 24.

Miller, William C. *Color photography in astronomy*, Publications of the Astronomical Society of the Pacific vol. 74, no. 441 (December 1962), p. 457.

Moore, Patrick. *Il guinness dell'astronomia*, Milano, Rizzoli, 1990.

Murray, Bruce C. *Mercurio*, "Le Scienze" no. 91 (March 1976), p. 40.

Newhall, Beaumont. *L'immagine latente: storia dell'invenzione della fotografia*, Bologna, Zanichelli, 1969.

Osterbrock, Donald E.; Gwinn, Joel A.; Brashear, Ronald S. *Edwin Hubble e l'espansione dell'universo*, "Le Scienze" no. 301 (September 1993), p. 70.

Paresce, Francesco. *Faint Object Camera: meglio non si può*, "L'Astronomia" no. 143 (May 1994), p. 16.

Paresce, Francesco. *SN 1987A: nuove sorprese*, "L'Astronomia" no. 146 (August-September 1994), p. 4.

Pierce, Michael J., et al. *The Hubble constant and Virgo cluster distance from observations of cepheid variables*, "Nature" vol. 371, no. 6496 (29 September 1994), p. 385.

Primmerman, Charles A., et al. *Compensation of atmospheric optical distortion using a synthetic beacon*, "Nature" vol. 353, no. 6340 (12 September 1991), p. 141.

Rees, Martin J. *Buchi neri al centro delle galassie*, "Le Scienze" no. 269 (January 1991), p. 24.

Romer, Grant B.; Delamoir, Jeannette. *Le prime fotografie a colori*, "Le Scienze" no. 258 (February 1990), p. 44.

Rosino, Leonida. *Le stelle variabili*, Roma, Curcio, 1988.

Sandberg, Thomas. *Gli albori della fotografia astronomica*, "L'Astronomia" no. 94 (December 1989), p. 58.

Schorn, Ronald A. *The extragalactic zoo I, II, III & IV*, "Sky & Telescope" vol. 75, nos. 1 and 4, pp. 23 e 376, vol. 76, nos. 1 and 4, pp. 36 & 344 (January, April, July and October 1988).

Sinnott, Roger W.; Nyren, Kari. *S & T's guide to the world's largest telescopes*, "Sky & Telescope" vol. 86, no. 1 (July 1993), p. 27.

Soker, Noam. *Le nebulose planetarie*, "Le Scienze" no. 287 (July 1992), p. 34.

Staehle, Robert L.; Terrile, Richard J.; Weinstein, Stacy S. *To Pluto by way of a postage stamp*, "The Planetary Report" vol. 14, no. 5 (September-October 1994), p. 4.

Stephenson, Richard F. *Macchie solari prima del telescopio*, "L'Astronomia" no. 113 (August-September 1991), p. 18.

Stern, Alano. *Where has Pluto's family gone?* "Astronomy" vol. 20, no. 9 (September 1992), p. 40.

Stone, Edward; Bane, Dono. *The Voyager flight to Jupiter and Saturn*, Pasadena, NASA-JPL, 1982.

Strom, Robert G. *"Guest stars", sample completeness and the local supernova rate*, "Astronomy and Astrophysics" vol. 288 (3 September 1994), L1.

Tanga, Paolo. *La cometa si è tuffata*, "L'Astronomia" no. 147 (October 1994), p. 12.

Thomson, Robert C.; Mackay, Craig D.; Wright, Alan E. *Internal structure and polarization of the optical jet of the quasar 3C 273*, "Nature" vol. 365, no. 6442 (9 September 1993), p. 133.

Thorsett, S.E. *Identification of the pulsar PSR 1509-58 with the "guest star" of AD 185*, "Nature" vol. 356, no. 6371 (23 April 1992), p. 690.

Trimble, Virginia; Parker, Samantha. *Meet the Milky Way*, "Sky & Telescope" vol. 89, no. 1 (January 1995), p. 26.

Van den Bergh, Sidney; Hesser, James E. *La formazione della Via Lattea*, "Le Scienze" no. 295 (March 1993), p. 36.

Vanin, Gabriele. *Il Voyager 2 incontra Nettuno*, "Astronomia UAI" no. 4 (July-August 1989), p. 3.

Vanin, Gabriele. *Il cielo dalla storia al futuro: un'introduzione all'astronomia moderna*, Galliera Veneta, Biroma, 1990.

Vanin, Gabriele. *Il presente e il futuro dei telescopi ottici*, "Astronomia UAI" no. 3 (May-June 1991), p. 8.

Vanin, Gabriele. *Le missioni interplanetarie degli anni 90*, "Astronomia UAI" no. 4 (July-August 1991), p. 11.

Vanin, Gabriele. *Le Pleiadi e la leggenda dell'Atlantide perduta*, "L'Astronomia" no. 129 (February 1993), p. 22.

Vanin, Gabriele. *Stelle cadenti*, Galliera Veneta, Biroma, 1994.

Vanin, Gabriele. *Le grandi comete*, "L'Astronomia" no. 142 (April 1994), p. 26.

*Viking: the exploration of Mars*, Pasadena, NASA-JPL, 1984.

*Voyager: the grandest tour*, Pasadena, NASA-JPL, 1991.

Weissman, Paul. *Bodies at the brink*, "The Planetary Report" vol. 14, no. 1 (January-February 1994), p. 4.

West, Richard M. *Europe's astronomy machine*, "Sky & Telescope" vol. 75, no. 5 (May 1988), p. 471.

Woosley, Stan; Weaver, Tom. *La grande supernova 1987A*, "Le Scienze" no. 254 (October 1989), p. 20.

Wray, James D. *The color atlas of galaxies*, New York, Cambridge University Press, 1988.

Yeomans, Donald. *Comets: a chronological history of observation, science, myth, and folklore*, New York, Wiley, 1991.

Zaritsky, Dennis; Rix, Hans-Walter; Rieke, Marcia. *Inner spiral structure of the galaxy M51*, "Nature" vol. 364, no. 6435 (22 July 1993), p. 313.

## INDEX OF ILLUSTRATIONS